AF531665

Global Environmental Law, Policy and Action Plan Series

Climate, Forest, Biodiversity and Desert

Global Environmental Law, Policy and Action Plan Series

Climate, Forest, Biodiversity and Desert

Edited by
Dr. Prabhas Chandra Sinha

2006

SBS Publishers & Distributors Pvt. Ltd.
New Delhi

ISBN : 81-89741-32-2

First Published in India in 2006

Published by:
SBS PUBLISHERS & DISTRIBUTORS PVT. LTD.
2/9, Ground Floor, Ansari Road, Darya Ganj,
New Delhi - 110002, INDIA
Tel: 23289119, 41563911
Email: mail@sbspublishers.com

Printed at
Chaman Enterprises
1603, Pataudi House,Daryaganj
New Delhi 110002

Preface

This book titled "Climate, Forest, Biodiversity and Desert" is an attempt to address the most important environment issues of current period. The global laws, policies and action plans on climate change are presented. The text of United Nations Framework Convention on Climate Change and its present Status of Ratification is given. Protecting the forests and forestry around the world has become a big challenge. This book presents the report of the United Nations Conference on Environment and Development in this regard. Global guidelines on biodiversity and biosafety are provided in detail. The full text of Convention on Biological Diversity and the current list of consultative parties to CBD are provided herein. In addition, the Cartagena Protocol on Biosafety is dealt in detail. The global law, policy and action plan to combat desertification is covered in totality. The text of United Nations Convention to Combat Desertification (UNCCD) applicable mainly in countries experiencing serious drought and/or desertification, particularly in Africa is presented. In addition, this book also covers the global strategy for conservation and management of resources for environment and development, *i.e.* the action plan as per Agenda 21 for: protection of the atmosphere; integrated approach to the planning and management of land resources; combating deforestation; managing fragile ecosystem such as combating desertification and drought; managing fragile ecosystem such as sustainable mountain development; promoting sustainable agriculture and rural development; conservation of biological diversity; and environmentally sound management of biotechnology. This book reflects in this regard about the Earth Summit, held in 1992, which addressed issues and challenges related to sustainable development at the global level. Delegates adopted Agenda 21, a plan for achieving sustainable development in the 21st Century. Since UNCED, significant progress has been made in the development of legislation, agreements and programmes of action at the international level with respect to global climate, forest, biodiversity and desert. The objective of this book is to provide readers/students with an integrated understanding of the global environmental regimes for climate change, forest protection, biodiversity conservation, preserving desert ecosystems.

Dr. Prabhas Chandra Sinha

Contents

Acronyms

ABC	Atmospheric Brown Cloud
ACC	Administrative Committee on Coordination (United Nations)
ACOMR	Advisory Committee on Oceanic Meteorological Research
ANWR	Arctic National Wildlife Refuge
AZE	Alliance for Zero Extinction
BOD	Biological Oxygen Demand
CBD	Convention on Biological Diversity
CCAMLR	Convention on the Conservation of Antarctic Marine Living Resources
CDIAC	Carbon Dioxide Information Analysis Center
CFCs	Chlorofluorocarbons
CHP	Combined Heat and Power
CIAP	Interamerican Committee of the Alliance for Progress
CMS	Convention on Migratory Species
CNG	Compressed Natural Gas
CO	Carbon Monoxide
DALY	Disability-Adjusted Life Years
DME	Dimethyl Ether
EAIS	East Antarctic Ice Sheet
EANET	Acid Deposition Monitoring Network East Asia
ECE	Economic Commission for Europe (United Nations)
ECLA	Economic Commission for Latin America (United Nations)
EEA	European Environment Agency
ENEA	European Nuclear Energy Agency
ETS	Emissions Trading Scheme (of the EU)
EU	European Union
EVI	Environmental Vulnerability Index

FAO	Food and Agriculture Organization of the United Nations
GARP	Global Atmospheric Research Programme
GATT	General Agreement on Tariffs and Trade
GCC	Gulf Cooperation Council
GDP	Gross Domestic Product
GEF	Global Environment Facility
GEMS	Global Environment Monitoring System
GEO	Global Environment Outlook (of UNEP)
GHG	Greenhouse Gas
GIPME	Global Investigation of Pollution in the Marine Environment
GM	Genetically Modified
GRASP	Great Apes Survival Project
GRID	Global Resource Information Database
HCFCs	Cydrochloroflourocarbons
HEVs	Hybrid Electric Vehicles
HIPC	Highly Indebted Poor Countries
IAEA	International Atomic Energy Agency
IARC	International Agency for Research on Cancer
IATA	International Air Transport Association
IBP	International Biological Programme
ICAO	International Civil Aviation Organization
ICC	International Computing Centre
ICE	International Centre for the Environment
ICES	International Council for the Exploration of the Sea
ICSPRO	Inter-Secretariat Committee on Scientific Problems Relating to Oceanography
ICSU	International Council of Scientific Unions
IEA	International Energy Agency
IGOSS	Integrated Global Ocean Station System
IHD	International Hydrological Decade
IIASA	International Institute for Applied Systems Analysis
ILO	International Labour Organisation
IMO	International Maritime Organization
INDOEX	Indian Ocean Experiment
IOC	Intergovernmental Oceanographic Commission
IPCC	Intergovernmental Panel on Climate Change
IPHE	International Partnership for the Hydrogen Economy
ISO	International Organization for Standardization

ISSS	International Society of Soil Science
ITU	International Telecommunication Union
IUCN	International Union for Conservation of Nature and Natural Resources
IUCN	World Conservation Union
IUFRO	International Union of Forestry Research Organization
IUU	Illegal, Unregulated and Unreported (fishing)
IVIC	Institute of Scientific Research, Caracas
IWG	Intergovernmental Working Group
IWP	Indicative World Plan
KISR	Kuwait Institute for Scientific Research
KOE	Kilogram of Oil Equivalent
LPG	Liquefied Petroleum Gas
MA	Millennium Ecosystem Assessment
MAB	Man and the Biosphere Programme (UNESCO)
MAP	Madre de Dios (Peru), Acre (Brazil) and Pando (Bolivia)
MDGs	Millennium Development Goals
MEA	Multilateral Environmental Agreement
MeBr	Methyl Bromide
NASA	National Aeronautics and Space Administration (of the United States)
NEPAD	New Partnership for Africa's Development
NO	Nitric Oxide
NO_2	Nitrogen Dioxide, Nitrites
NO_3	Nitrates
NOAA	National Oceanic and Atmospheric Administration (of the United States)
NOx	Nitrogen Oxides
NPRA	National Petroleum Reserve in Alaska
O_3	Ozone
ODP	Ozone Depleting Potential
ODS	Ozone Depleting Substance
PAHs	Polyaromatic Hydrocarbons
PCP	Pentachlorophenol
PIC	Prior Informed Consent
PM	Particulate Matter
POPs	Persistent Organic Pollutants
PPP	Purchasing Power Parity
PV	Photovoltaic

RLIs	Red List Indices
SCOPE	Scientific Committee On Problems of the Environment
SCOR	Scientific Committee on Ocean Research
SIDS	Small Island Developing State(s)
SO_2	Sulphur Dioxide
SPM	Suspended Particulate Matter
Stockholm Convention	Stockholm Convention on Persistent Organic Pollutants
UAE	United Arab Emirates
UK	United Kingdom
UN	United Nations
UNDP	United Nations Development Programme
UNECE	United Nations Economic Commission for Europe
UNEP	United Nations Environment Programme
UNESCO	United Nations Educational, Scientific and Cultural Organization
UNESOB	United Nations Economic and Social Office in Beirut
UNFCCC	United Nations Framework Convention on Climate Change
UNFPA	United Nations Fund for Population Activities
UNICEF	United Nations Children's Fund
UNIDO	United Nations Industrial Development organization
UNISIST	World Science Information System
UNITAR	United Nations Institute for Training and Research
UNRWA	United Nations Relief and Works Agency for Palestine Refugees
UNSCEAR	United Nations Scientific Committee on the Effects of Atomic Radiation
US	United States
USEPA	United States Environmental Protection Agency
USGS	United States Geological Service
UV	Ultraviolet
WAIS	West Antarctic Ice Sheet
WEI	World Environment Institute
WFP	World Food Programme
WHO	World Health Organization
WMO	World Meteorological Organization
WWF	World Wide Fund for Nature
WWW	World Weather Watch

Glossary of Terms

Acid Rain or Acid Deposition
A complex chemical and atmospheric phenomenon that occurs when emissions of sulphur and nitrogen compounds and other substances are transformed by chemical processes in the atmosphere, often far from the original sources, and then deposited on earth in either a wet or dry form. The wet forms, popularly called acid rain, can fall as rain, snow or fog. The dry forms are acidic gases or particulates.

Active Solar Energy
The advanced use of sunshine through solar photovoltaic or solar thermal conversion technologies.

Air Quality Standards
The level of pollutants prescribed by regulations that may not be exceeded during a specified time in a defined area.

Agricultural Residues
By-products of the agricultural production system, including straws, husks, shells stalks and animal dung, with a large number of uses, including an energy resource. Residues can be divided into two groups: crop residues, which remain on the field after harvest, for example, cotton stalks; and agro-processing residues, produced off-field at a central production sight, for example, rice husk.

Alternative Energy System
A phrase used to describe new energy technologies and production processes that are seen as alternatives for traditional and conventional energy sources.

Anaerobic Digestion
The natural process of decomposition of organic material by bacteria under anaerobic conditions (i.e. in the absence of oxygen), a by-product of which is biogas.

Animate Energy (or power)
Utilisation of human and animal muscles to do work.

Assimilation
The ability of a body of water to purify itself of pollutants.

Base Load
The average minimum demand on an electricity system, usually measured over a 24 hour cycle.

Basel Convention
The Basel Convention on the Control of Trans-boundary Movements of Hazardous Wastes and their Disposal (1989) aims to control the trans boundary movement and disposal of hazardous wastes.

Biochemical Oxygen Demand (BOD)
A measure of the amount of oxygen consumed in the biological processes that break down organic matter in water. The greater the BOD, the greater the pollution.

Biodegradability
The ability to break down or decompose rapidly under natural conditions and processes.

Bio-diesel
Modified (trans-esterified) vegetable oils, such as soya bean, palm oil, jatropha, which have similar chemical properties to diesel and therefore can be used as a direct substitute.

Biodiversity
Biological diversity, the variety of species within a given area.

Biogas
A mixture of gases consisting of around 60 to 70% of methane (depending on conditions) produced by the process of anaerobic digestion in a closed container (digester). Biogas burns with similar properties to natural gas.

Biological Magnification
Refers to the process whereby certain substances such as pesticides or heavy metals move up the food chain, work their way into a river or lake and are eaten by large birds, animals or humans. The substances become concentrated in tissues or internal organs as they move up the chain.

Biological Oxidation

The way bacteria and micro-organisms feed on and decompose complex organic materials. Used in self-purification of water bodies and in activated sludge waste-water treatment.

Biological Treatment

A treatment technology that uses bacteria to consume waste. This treatment breaks down organic materials.

Biomass

A generic term to describe all forms of organic matter including wood, agricultural waste, animal dung and human waste. The term can be extended to include derivatives, such as ethanol, which can be used as fuels. (The scientific definition is the total dry organic matter or stored energy content of living organisms in a given area.)

Biosphere

The portion of the earth that supports life, including the surface waters and the air. Similar to ecosphere.

Biotechnology

Techniques that use living organisms or parts of organisms to produce a variety of products, from medicines to industrial enzymes, to improve plants or animals or to develop micro-organisms for specific uses such as removing toxins from bodies of water or killing pests.

Briquetting

Densification of loose organic material, such as rice husk, sawdust, coffee husks, to improve fuel characteristics including handling and combustion properties.

Cadmium

Toxic heavy metal used mainly for metal plating and as a plastic pigment. Significant by-product in zinc smelting and of concern in phosphate manufacture.

Capacity Factor

The actual output of a technology divided by the theoretical maximum output of the technology operating at peak design resource levels.

Capital Cost

The investment in plant and equipment, including construction costs but not operation, maintenance or energy costs.

Carbon Cycle
The circulation of carbon in the biosphere. Carbon is an essential part of the molecules that make up all living cells. In the atmosphere it circulates as carbon dioxide, which is released by respiration, combustion and decay and fixed in complex carbon compounds during photosynthesis in plants and certain micro-organisms.

Carbon Dioxide (CO_2)
A colourless, odourless, non-poisonous gas that results from respiration, combustion and decay and is normally present in ambient air.

Carbonisation
The heating of biomass in the absence of air to remove the volatile component, leaving a solid high carbon product, charcoal.

Catalytic Converter
An air pollution abatement device that removes pollutants from motor vehicle exhaust, either by oxidising them into carbon dioxide and water or reducing them to nitrogen and oxygen.

Chlorofluorocarbons (CFCs)
A family of inert, non-toxic and easily liquefied chemicals used in refrigeration, air conditioning, packaging and insulation or as solvents and aerosol propellants. Because CFCs are not destroyed in the lower atmosphere they drift into the upper atmosphere, where their chlorine components destroy ozone.

Cleaner Production
A concept of industrial production that minimises all environmental impacts through careful management of resource use, of product design and use, systematic waste avoidance and management of residuals, safe working practices and industrial safety. Sometimes called pollution prevention or waste minimisation.

Clean Technologies
Production processes or equipment with a low rate of waste production. Treatment or recycling plants are not classed as clean technologies.

Combined Heat and Power (CHP) Generation
Simultaneous production of both heat and power. The difference with co-generation is that the generation of heat and power may be done as parallel processes, which results in a lower overall efficiency.

Combustion
Chemical reaction between a fuel and oxygen which usually takes place in air. More commonly known as burning. The products are carbon dioxide and water with the release of heat energy.

Commercial Energy
Energy traded in the market for a monetary price, usually conventional energy, such as coal or oil, but also traditional energy, such as fuelwood, which is traded in urban and semi-urban areas in many developing countries.

Coppicing
Technique of utilising the ability of some trees species to re-generate from the stump after a cutting.

Cradle-to-grave
Term used to imply the whole life cycle of a product, from raw material to final disposal.

Decentralised Energy Systems
Small scale energy producing facilities deigned to meet local demand only.

Demand Side Management
The planning, implementation and monitoring of activities designed to encourage consumers to modify their pattern of energy use.

Dendropower
The conversion of wood energy to electricity or shaft power.

Dilution Ratio
The relationship between the volume of water in a stream and the volume of incoming water. If affects the ability of the stream to assimilate waste.

Dioxin
Any of a family of compounds known chemically as dibénzo-*p*-dioxins. They are of concern because of their potential toxicity and contamination in commercial products. Tests on laboratory animals indicate that dioxins are among the more toxic man-made chemicals known.

Disposal
Final placement or destruction of toxic, radioactive or other wastes; surplus or banned pesticides or other chemicals; polluted soils; and drums containing hazardous materials from removal actions or accidental releases. Disposal may be accomplished through use of approved secure landfills, surface impoundments, land farming, deep well injection, ocean dumping or incineration.

Dissolved Oxygen (DO)
The oxygen freely available in water. Dissolved oxygen is vital to fish and other aquatic life and for the prevention of odours. Traditionally, its level has been accepted as the single most important indicator of a water body's ability to support desirable aquatic life. Secondary and advanced waste treatment are generally designed to protect DO in waste-receiving waters.

Dump
A site used to dispose of solid wastes without environmental controls.

Eco-efficiency
Maximization of industrial output from a given level of resource input, thus ensuring waste minimisation and appropriate use of human, renewable and non-renewable resources.

Ecology
The relationship of living things to one another and their environment, or the study of such relationships.

Ecologically Sustainable Industrial Development (ESID)
Patterns of industrialisation that enhance the contribution of industry to economic and social benefits for present and future generations without impairing basic ecological processes.

Ecosystem
The interacting system of a biological community and its non-living environmental surroundings.

End-of-pipe Treatment (abatement)
Treating pollutants at the end of a process (by, for example, filters, catalysts and scrubbers) instead of preventing their occurrence.

End-use
The final use of energy at the level of the user, for example, lighting, cooking, space-heating and motive power.

Energy
Capacity (or ability) to do work.

Environment
The sum of all external conditions including physical and social factors, affecting the life, development and survival of an organism.

Environmental Impact Assessment
An analysis to determine whether an action would significantly affect the environment.

Eutrophication
The slow ageing process in which a lake, estuary or bay becomes a bog or marsh and eventually disappears. During the later stages, or eutrophication, the water body is choked by overabundant plant life as the amounts of nutritive compounds such as nitrogen and phosphorus increase. Human activities can accelerate the process.

Fossil Fuels
The non-renewable energy resources of coal, petroleum or natural gas or any fuel derived from them. Exploitation of new reserves has a long development period.

Fuel Wood
Trunks and branches of trees as well as residues from wood processing industries and scrapwood which are used as an energy source, largely through direct combustion.

Gasification
Conversion of solid fuels (biomass and coal) at high temperatures in the absence of air to give combustible gases.

Generating Capacity
The capacity of a power plant to generate electricity, often expressed in watts-electric (e.g. MWe).

Global Environment Monitoring System
Global Environment Monitoring System, managed by UNEP. Makes comprehensive assessments of major environmental issues and thus provides the scientific data needed for the rational management or natural resources and the environment, provides early warning of environmental changes by analysing monitoring data.

Global Warming
The consequences of the greenhouse effect, caused by rising concentrations of greenhouse gases. The suspicion is that global warming will disrupt weather and climate patterns. It could lead to drought in some areas and flooding in others. One of the most serious environmental problems facing the world.

Good Housekeeping
Efficient management of the property and equipment of an institution or organisation. In the context of Cleaner Production, it often refers to the procedures applied in the operation of a production process.

Greenhouse Effect
The warming of the earth's atmosphere, caused by a build-up of carbon dioxide or other trace gases. It is believed by many scientists that this build-up allows light from the sun's ray to heat the earth but prevents a counterbalancing loss of heat.

Greenhouse Gases
The gases, such as carbon dioxide, water vapour, methane, nitrous oxides and CFCs, that trap the sun's heat in the lower atmosphere and prevent it from escaping into space. Major source of increased concentration in the atmosphere is the combustion of fossil fuels.

Groundwater
The supply of fresh water found beneath the earth's surface (usually in aquifers) that is often used for supplying wells and springs. Because groundwater is an important source of drinking water, there is growing concern about areas where agricultural or industrial pollutants or substances from leaking underground storage tanks are contaminating groundwater.

Halons
Bromine-containing compounds with long atmospheric lifetimes whose breakdown in the stratosphere can cause depletion of ozone. Halons are used in fire-fighting.

Hazardous Waste
By-products of society that can pose a substantial hazard to human health or the environment when improperly managed. Characterised by at least one of the following: ignitability, corrosivity, reactivity or toxicity.

Heavy Metals
Metallic elements with high atomic weights, e.g. mercury, chromium, cadmium, arsenic and lead. They can damage living organisms at low concentrations and tend to accumulate in the food chain.

Hydrocarbon (HC)
Chemical compounds consisting entirely of carbon and hydrogen.

Landfills
Sanitary landfills are land disposal sites for non-hazardous solid wastes, where the waste is spread in layers, compacted to the smallest practical volume and cover material applied at the end of each operating day.

Leachate
A liquid produced when water collects contaminants as it trickles through wastes, agricultural pesticides of fertilisers. Leaching may occur in farming areas, feedlots or landfills and may result in hazardous substances entering surface water, groundwater or soil.

Life Cycle Cost
The cost of a good or service over its entire life cycle.

Mercury
A heavy metal that can accumulate in the environment and is highly toxic if breathed or swallowed.

Minimisation
Actions to avoid, reduce or in other ways diminish the hazards of wastes at their source. Recycling is, strictly speaking, not a minimisation technique but is often included in such programmes for practical reasons.

Monoculture
The exclusive cultivation of a single species, a common practice in modern agriculture and forestry.

Montreal Protocol
The Montreal Protocol on Substances that Deplete the Ozone Layer, adopted 16 September 1987, sets limits for the production and consumption of damaging CFCs and halons.

Nitrate
A compound containing nitrogen that can exist in the atmosphere or as a dissolved gas in water and that can have harmful effects on humans and animals. Nitrates in water can cause severe illness in infants and cows.

Non-commercial Energy
Energy which does not have a monetary price, usually refers to fuelwood, agricultural residues or animal waste, which are self collected or traded through barter.

Non-point Source
Pollution sources that are diffuse and do not have a single point of origin or are not introduced into a receiving stream from a specific outlet. The pollutants are generally carried off the land by storm-water run-off. The commonly used categories for non-point sources are agriculture, forestry, urban areas, mining, construction, dams and channels, land disposal and salt-water intrusion.

Non-renewable Energy
Any form of primary energy, the supply of which is finite and hence its use depletes the existing stock. It generally refers to fossil fuels.

Non-woody Biomass
Stalks, leaves, grass, animal and human waste.

NOx
Chemical formula that stands for all the oxides of nitrogen, mainly NO_2, but also N_2O, NO, N_2O_3, N_2O_4 and NO_3, which is unstable.

Nutrient
Any substance assimilated by living organisms that promotes growth. The term is generally applied to nitrogen and phosphorus in waste water but is also applied to other essential and trace elements.

Off-site Facility
A hazardous waste treatment, storage or disposal area that is located away from the generating site.

Ozone (O_3)
Found in the lower layers of the atmosphere, the stratosphere and the troposphere. In the stratosphere (the atmospheric layer beginning 7-10 miles above the earth's surface), ozone is a form of

oxygen found naturally which provides a protective layer, shielding the earth from ultraviolet radiation's harmful health effects on humans and the environment. In the troposphere (the layer extending up 7-10 miles above the earth's surface), ozone is a chemical oxidant and a major component of photochemical smog. Ozone can seriously affect the human respiratory system and is one of the most prevalent and widespread pollutants. Ozone in the troposphere is produced through complex chemical reactions between nitrogen oxides, which are among the primary pollutants emitted by combustion sources, hydrocarbons, which are released into the atmosphere by the combustion, handling and processing of petroleum products, and sunlight.

Ozone Depletion
Destruction of the stratospheric ozone layer, which shields the earth from ultraviolet radiation harmful to life. This destruction of ozone is caused by certain chlorine- and/or bromine-containing compounds (chlorofluorocarbons or halons), which break down when they reach the stratosphere and catalyse the destruction of ozone molecules.

Passive Solar Energy
The direct use of natural sunshine for lighting, heating or convective ventilation in buildings in order to replace conventional conversion technologies.

Peak Load
The average maximum power demand on an energy system, especially for electricity generation, over a period of time, usually a 24 hour cycle.

Phenols
Organic compounds that are by-products of petroleum refining, tanning and the manufacturing of textiles, dyes and resins. Low concentrations cause taste and odour problems in water; higher concentrations can kill aquatic, animal and human life.

Phosphorus
An essential chemical food element that can contribute to the eutrophication of lakes and other water bodies. Increased phosphorus levels result from discharge of phosphorus containing materials into surface waters.

Photochemical Smog
Air pollutant formed by the action of sunlight on oxides of nitrogen and hydrocarbons.

Pollution
Generally, the presence of matter or energy whose nature, location or quantity produces undesired environmental effects.

Primary Energy Sources
Any natural energy source available in nature which can be transformed into a useful energy form, such as coal, solar, biomass.

PVC
A tough, environmentally indestructible plastic (chemical name: polyvinyl chloride) that releases hydrochloric acid when burned.

Recycling
The process of minimising the generation of waste by recovering useable products that might otherwise become waste. Examples are the recycling of aluminium cans, waste paper and bottles.

Reserves
The portion of a resource base that is proven to exist and can be economically recovered, that is, the value of the product exceeds the production and transportation costs.

Residual
A pollutant remaining in the environment after a natural or technological process has taken place, e.g. the sludge remaining after initial waste-water treatment or particulates remaining in air after the air passes through a scrubbing or other pollutant removal process.

Resources
The total existing stock of a resource, including discovered and not yet discovered portions, regardless of the economic feasibility of recovering the resource.

Risk Assessment
The qualitative and quantitative evaluation performed in an effort to define the risk posed to human health and/or the environment by the presence or potential presence and/or use of specific pollutants.

Rural Electrification
The extension of supply of electricity to the rural areas, conventionally by extending the national grid based on centralised generating power plants but now taken to include small scale decentralised generating systems.

Scrubber
An air pollution control device that uses a spray of water or reactant or a dry process to trap pollutants in emissions.

Secondary Energy
Energy in a form ready for transport or transmission.

Sink
In air pollution, the receiving area for material removed from the atmosphere, e.g. by virtue of photosynthesis, plants are sinks for carbon dioxide.

Solar Photovoltaics
Technologies which convert sunshine into electricity using solar cells (photosensitive silicon cells).

Solar Thermal Energy
Energy derived from converting sunshine into direct heating applications, such as water heating, or to produce electricity in a steam generating plant.

Solvents
Liquids that dissolve other substances. Chemical solvents are used widely in industry. They are used by pharmaceutical makers to extract active substances; by electronics manufacturers to wash circuit boards; by paint-makers to aid drying. Most solvents can cause air and water pollution and can be a health hazard.

Sulphur Dioxide (SO_2)
A colourless, irritating pungent gas formed when sulphur burns in air, one of the main air pollutants that contribute to acid rain and smog. Comes from the combustion of the sulphur present in most fossil fuels.

Suspended Particulate Matter (SPM)
Fine liquid or solid particles such as dust, smoke, mist, fumes or smog, found in air or emissions.

Sustainable Development
Development that meets present needs without compromising the ability of future generations to meet their own needs. Necessarily based on limited data due to our current inability to forecast accurately 50-100 years ahead.

Tradeable Permits
Market mechanism for controlling pollution; it entails issuing permits to pollute up to fixed limits and grants the right to sell unused portions of the permits.

Traditional Energy
Forms of energy derived from locally available biomass, animate power and other renewable energy sources using rudimentary production processes and technologies.

Tree Production Rate
Annual increment of standing stock per unit area.

Waste Minimisation
The reduction of waste by changing materials, processes or on-site disposals arrangements in a way that is profitable for the enterprise and the environment. Also called waste reduction.

Water Quality Standards
Ambient standards for water bodies. The standards address the use of the water body and set water quality criteria that must be met to protect the designated use or uses.

Wind Farm
A group (or array) of wind turbines connected to a grid for producing electricity.

Woody Biomass
Stems, branches, shrubs, hedges, twigs and residues of wood processing.

Chapter 1

Global Laws, Policies and Action Plans on Climate Change

UNITED NATIONS FRAMEWORK CONVENTION ON CLIMATE CHANGE

The Parties to this Convention.

Acknowledging that change in the Earth's climate and its adverse effects are a common concern of humankind.

Concerned that human activities have been substantially increasing the atmospheric concentrations of greenhouse gases, that these increases enhance the natural greenhouse effect, and that this will result on average in an additional warming of the Earth's surface and atmosphere and may adversely affect natural ecosystems and humankind.

Noting that the largest share of historical and current global emissions of greenhouse gases has originated in developed countries, that per capita emissions in developing countries are still relatively low and that the share of global emissions originating in developing countries will grow to meet their social and development needs.

Aware of the role and importance in terrestrial and marine ecosystems of sinks and reservoirs of greenhouse gases.

Noting that there are many uncertainties in predictions of climate change, particularly with regard to the timing, magnitude and regional patterns thereof.

Acknowledging that the global nature of climate change calls for the widest possible cooperation by all countries and their participation in an effective and appropriate international response, in accordance with their common but differentiated responsibilities and respective capabilities and their social and economic conditions.

Recalling the pertinent provisions of the Declaration of the United Nations Conference on the Human Environment, adopted at Stockholm on 16 June 1972.

Recalling also that States have, in accordance with the Charter of the United Nations and the principles of international law, the sovereign right to exploit their own resources pursuant to their own environmental and developmental policies, and the responsibility to ensure that activities within their jurisdiction or control do not cause damage to the environment of other States or of areas beyond the limits of national jurisdiction.

Reaffirming the principle of sovereignty of States in international cooperation to address climate change.

Recognizing that States should enact effective environmental legislation, that environmental standards, management objectives and priorities should reflect the environmental and developmental context to which they apply, and that standards applied by some countries may be inappropriate and of unwarranted economic and social cost to other countries, in particular developing countries.

Recalling the provisions of General Assembly resolution 44/228 of 22 December 1989 on the United Nations Conference on Environment and Development, and resolutions 43/53 of 6 December 1988, 44/207 of 22 December 1989, 45/212 of 21 December 1990 and 46/169 of 19 December 1991 on protection of global climate for present and future generations of mankind.

Recalling also the provisions of General Assembly resolution 44/206 of 22 December 1989 on the possible adverse effects of sea-level rise on islands and coastal areas, particularly low-lying coastal areas and the pertinent provisions of General Assembly resolution 44/172 of 19 December 1989 on the implementation of the Plan of Action to Combat Desertification.

Recalling further the Vienna Convention for the Protection of the Ozone Layer, 1985, and the Montreal Protocol on Substances that Deplete the Ozone Layer, 1987, as adjusted and amended on 29 June 1990.

Noting the Ministerial Declaration of the Second World Climate Conference adopted on 7 November 1990.

Conscious of the valuable analytical work being conducted by many States on climate change and of the important contributions of the World Meteorological Organization, the United Nations Environment Programme and other organs, organizations and bodies of the United Nations system, as well as other international and intergovernmental bodies, to the exchange of results of scientific research and the coordination of research.

Recognizing that steps required to understand and address climate change will be environmentally, socially and economically most effective if they are based on relevant scientific, technical and economic considerations and continually re-evaluated in the light of new findings in these areas.

Recognizing that various actions to address climate change can be justified economically in their own right and can also help in solving other environmental problems.

Recognizing also the need for developed countries to take immediate action in a flexible manner on the basis of clear priorities, as a first step towards comprehensive response strategies at the global, national and, where agreed, regional levels that take into account all greenhouse gases, with due consideration of their relative contributions to the enhancement of the greenhouse effect.

Recognizing further that low-lying and other small island countries, countries with low-lying coastal, arid and semiarid areas or areas liable to floods, drought and desertification, and developing countries with fragile mountainous ecosystems are particularly vulnerable to the adverse effects of climate change.

Recognizing the special difficulties of those countries, especially developing countries, whose economies are particularly dependent on fossil fuel production, use and exportation, as a consequence of action taken on limiting greenhouse gas emissions.

Affirming that responses to climate change should be coordinated with social and economic development in an integrated manner with a view to avoiding adverse impacts on the latter, taking into full account the legitimate priority needs of developing countries for the achievement of sustained economic growth and the eradication of poverty.

Recognizing that all countries, especially developing countries, need access to resources required to achieve sustainable social and economic development and that, in order for developing countries to progress towards that goal, their energy consumption will need to grow taking into account the possibilities for achieving greater energy efficiency and for controlling greenhouse gas emissions in general, including through the application of new technologies on terms which make such an application economically and socially beneficial.

Determined to protect the climate system for present and future generations.

Have agreed as follows:

Article 1. Definitions—For the purposes of this Convention:

(1) "Adverse effects of climate change" means changes in the physical environment or biota resulting from climate change which have significant deleterious effects on the composition, resilience or productivity of natural and managed ecosystems or on the operation of socio-economic systems or on human health and welfare.

(2) "Climate change" means a change of climate which is attributed directly or indirectly to human activity that alters the composition of the global atmosphere and which is in addition to natural climate variability observed over comparable time periods.

(3) "Climate system" means the totality of the atmosphere, hydrosphere, biosphere and geosphere and their interactions.

(4) "Emissions" means the release of greenhouse gases and/or their precursors into the atmosphere over a specified area and period of time.

(5) "Greenhouse gases" means those gaseous constituents of the atmosphere, both natural and anthropogenic, that absorb and re-emit infrared radiation.

(6) "Regional economic integration organization" means an organization constituted by sovereign States of a given region which has competence in respect of matters governed by this Convention or its protocols and has been duly authorized, in accordance with its internal procedures, to sign, ratify, accept, approve or accede to the instruments concerned.

(7) "Reservoir" means a component or components of the climate system where a greenhouse gas or a precursor of a greenhouse gas is stored.

(8) "Sink" means any process, activity or mechanism which removes a greenhouse gas, an aerosol or a precursor of a greenhouse gas from the atmosphere.

(9) "Source" means any process or activity which releases a greenhouse gas, an aerosol or a precursor of a greenhouse gas into the atmosphere.

Article 2. Objective—The ultimate objective of this Convention and any related legal instruments that the Conference of the Parties may adopt is to achieve, in accordance with the relevant provisions of the Convention, stabilization of greenhouse gas concentrations in the atmosphere at a level that would prevent dangerous anthropogenic interference with the climate system. Such a level should be achieved within a time-frame sufficient to allow ecosystems to adapt naturally to climate change, to ensure that food production is not threatened and to enable economic development to proceed in a sustainable manner.

Article 3. Principles—In their actions to achieve the objective of the Convention and to implement its provisions, the Parties shall be guided, *inter alia,* by the following:

(1) The Parties should protect the climate system for the benefit of present and future generations of humankind, on the basis of equity and in accordance with their common but differentiated responsibilities and respective capabilities. Accordingly, the developed country Parties should take the lead in combating climate change and the adverse effects thereof.

(2) The specific needs and special circumstances of developing country Parties, especially those that are particularly vulnerable to the adverse effects of climate change, and of those Parties, especially developing country Parties, that would have to bear a disproportionate or abnormal burden under the Convention, should be given full consideration.

(3) The Parties should take precautionary measures to anticipate, prevent or minimize the causes of climate change and mitigate its adverse effects. Where there are threats of serious or irreversible damage, lack of full scientific certainty should not be used as a reason for postponing such measures, taking into account that policies and measures to deal with climate change should be cost-effective so as to ensure global benefits at the lowest possible cost. To achieve this, such policies and measures should take into account different socio-economic contexts, be comprehensive, cover all relevant sources, sinks and reservoirs of greenhouse gases and adaptation, and comprise all economic sectors. Efforts to address climate change may be carried out cooperatively by interested Parties.

(4) The Parties have a right to, and should, promote sustainable development. Policies and measures to protect the climate system

against human-induced change should be appropriate for the specific conditions of each Party and should be integrated with national development programmes, taking into account that economic development is essential for adopting measures to address climate change.

(5) The Parties should cooperate to promote a supportive and open international economic system that would lead to sustainable economic growth and development in all Parties, particularly developing country Parties, thus enabling them better to address the problems of climate change. Measures taken to combat climate change, including unilateral ones, should not constitute a means of arbitrary or unjustifiable discrimination or a disguised restriction on international trade.

Article 4. Commitments—(1) All Parties, taking into account their common but differentiated responsibilities and their specific national and regional development priorities, objectives and circumstances, shall:

(*a*) Develop, periodically update, publish and make available to the Conference of the Parties, in accordance with Article 12, national inventories of anthropogenic emissions by sources and removals by sinks of all greenhouse gases not controlled by the Montreal Protocol, using comparable methodologies to be agreed upon by the Conference of the Parties;

(*b*) Formulate, implement, publish and regularly update national and, where appropriate, regional programmes containing measures to mitigate climate change by addressing anthropogenic emissions by sources and removals by sinks of all greenhouse gases not controlled by the Montreal Protocol, and measures to facilitate adequate adaptation to climate change;

(*c*) Promote and cooperate in the development, application and diffusion, including transfer, of technologies, practices and processes that control, reduce or prevent anthropogenic emissions of greenhouse gases not controlled by the Montreal Protocol in all relevant sectors, including the energy, transport, industry, agriculture, forestry and waste management sectors;

(*d*) Promote sustainable management, and promote and cooperate in the conservation and enhancement, as appropriate, of sinks and reservoirs of all greenhouse gases not controlled by the Montreal Protocol, including biomass, forests and oceans as well as other terrestrial, coastal and marine ecosystems;

(*e*) Cooperate in preparing for adaptation to the impacts of climate change; develop and elaborate appropriate and integrated plans for coastal zone management, water resources and agriculture, and for the protection and rehabilitation of areas, particularly in Africa, affected by drought and desertification, as well as floods;

(*f*) Take climate change considerations into account, to the extent feasible, in their relevant social, economic and environmental policies and actions, and employ appropriate methods, for example impact assessments, formulated and determined nationally, with a view to minimizing adverse effects on the economy, on public health and on the quality of the environment, of projects or measures undertaken by them to mitigate or adapt to climate change;

(*g*) Promote and cooperate in scientific, technological, technical, socio-economic and other research, systematic observation and development of data archives related to the climate system and intended to further the understanding and to reduce or eliminate the remaining uncertainties regarding the causes, effects, magnitude and timing of climate change and the economic and social consequences of various response strategies;

(*h*) Promote and cooperate in the full, open and prompt exchange of relevant scientific, technological, technical, socio-economic and legal information related to the climate system and climate change, and to the economic and social consequences of various response strategies;

(*i*) Promote and cooperate in education, training and public awareness related to climate change and encourage the widest participation in this process, including that of non-governmental organizations; and

(*j*) Communicate to the Conference of the Parties information related to implementation, in accordance with Article 12.

(2) The developed country Parties and other Parties included in Annex I commit themselves specifically as provided for in the following:

(*a*) Each of these Parties shall adopt national1 policies and take corresponding measures on the mitigation of climate change, by limiting its anthropogenic emissions of greenhouse gases and protecting and enhancing its greenhouse gas sinks and reservoirs. These policies and measures will demonstrate that developed countries are taking the lead in modifying longer-term trends in anthropogenic emissions consistent with the objective of the Convention, recognizing that the return by the end of the present decade to earlier levels of anthropogenic emissions of carbon dioxide and other greenhouse gases not controlled by the Montreal Protocol would contribute to such modification, and taking into account the differences in these Parties' starting points and approaches, economic structures and resource bases, the need to maintain strong and sustainable economic growth, available technologies and other individual circumstances, as well as the need for equitable and appropriate contributions by each of these Parties to the global effort regarding that objective. These Parties may implement such policies and measures jointly with other Parties and may assist other Parties in contributing to the achievement of the objective of the Convention and, in particular, that of this subparagraph;

(*b*) In order to promote progress to this end, each of these Parties shall communicate, within six months of the entry into force of the Convention for it and periodically thereafter, and in accordance with Article 12, detailed information on its policies and measures referred to in subparagraph (*a*) above, as well as on its resulting projected anthropogenic emissions by sources and removals by sinks of greenhouse gases not controlled by the Montreal Protocol for the period referred to in subparagraph (*a*), with the aim of returning individually or jointly to their 1990 levels these anthropogenic emissions of carbon dioxide and other greenhouse gases not controlled by the Montreal Protocol. This information will be reviewed by the Conference of the Parties, at its first session and periodically thereafter, in accordance with Article 7;

(*c*) Calculations of emissions by sources and removals by sinks of greenhouse gases for the purposes of subparagraph (*b*) above should take into account the best available scientific knowledge, including of the effective capacity of sinks and the respective contributions of such gases to climate change. The Conference of the Parties shall consider and agree on methodologies for these calculations at its first session and review them regularly thereafter;

(*d*) The Conference of the Parties shall, at its first session, review the adequacy of subparagraphs (*a*) and (*b*) above. Such review shall be carried out in the light of the best available scientific information and assessment on climate change and its impacts, as well as relevant technical, social and economic information. Based on this review, the Conference of the Parties shall take appropriate action, which may include the adoption of amendments to the commitments in subparagraphs (*a*) and (*b*) above. The Conference of the Parties, at its first session, shall also take decisions regarding criteria for joint implementation as indicated in subparagraph (*a*) above. A second review of subparagraphs (*a*) and (*b*) shall take place not later than 31 December 1998, and thereafter at regular intervals determined by the Conference of the Parties, until the objective of the Convention is met;

(*e*) Each of these Parties shall:

(*i*) Coordinate as appropriate with other such Parties, relevant economic and administrative instruments developed to achieve the objective of the Convention; and

(*ii*) Identify and periodically review its own policies and practices which encourage activities that lead to greater levels of anthropogenic emissions of greenhouse gases not controlled by the Montreal Protocol than would otherwise occur;

(*f*) The Conference of the Parties shall review, not later than 31 December 1998, available information with a view to taking decisions regarding such amendments to the lists in Annexes I and II as may be appropriate, with the approval of the Party concerned;

(*g*) Any Party not included in Annex I may, in its instrument of ratification, acceptance, approval or accession, or at any time thereafter, notify the Depositary that it intends to be bound by subparagraphs (*a*) and (*b*) above. The Depositary shall inform the other signatories and Parties of any such notification.

(3) The developed country Parties and other developed Parties included in Annex II shall provide new and additional financial resources to meet the agreed full costs incurred by developing country Parties in complying with their obligations under Article 12, paragraph 1. They shall also provide such financial resources, including for the transfer of technology, needed by the developing country Parties to meet the agreed full incremental costs of implementing measures that are covered by paragraph 1 of this Article and that are agreed between a developing country Party and the international entity or entities referred to in Article 11, in accordance with that Article. The implementation of these commitments shall take into account the need for adequacy and predictability in the flow of funds and the importance of appropriate burden sharing among the developed country Parties.

(4) The developed country Parties and other developed Parties included in Annex II shall also assist the developing country Parties that are particularly vulnerable to the adverse effects of climate change in meeting costs of adaptation to those adverse effects.

(5) The developed country Parties and other developed Parties included in Annex II shall take all practicable steps to promote, facilitate and finance, as appropriate, the transfer of, or access to, environmentally sound technologies and knowhow to other Parties, particularly developing country Parties, to enable them to implement the provisions of the Convention. In this process, the developed country Parties shall support the development and enhancement of endogenous capacities and technologies of developing country Parties. Other Parties and organizations in a position to do so may also assist in facilitating the transfer of such technologies.

(6) In the implementation of their commitments under paragraph 2 above, a certain degree of flexibility shall be allowed by the Conference of the Parties to the Parties included in Annex I undergoing the process of transition to a market economy, in order to enhance the ability of these Parties to address climate change, including with regard to the historical level of anthropogenic emissions of greenhouse gases not controlled by the Montreal Protocol chosen as a reference.

(7) The extent to which developing country Parties will effectively implement their commitments under the Convention will depend on the effective implementation by developed country Parties of their commitments under the Convention related to financial resources and transfer of technology and will take fully into account that economic and social development and poverty eradication are the first and overriding priorities of the developing country Parties.

(8) In the implementation of the commitments in this Article, the Parties shall give full consideration to what actions are necessary under the Convention, including actions related to funding, insurance and the transfer of technology, to meet the specific needs and concerns of developing country Parties arising from the adverse effects of climate change and/or the impact of the implementation of response measures, especially on:

(*a*) Small island countries;
(*b*) Countries with low-lying coastal areas;
(*c*) Countries with arid and semi-arid areas, forested areas and areas liable to forest decay;
(*d*) Countries with areas prone to natural disasters;
(*e*) Countries with areas liable to drought and desertification;
(*f*) Countries with areas of high urban atmospheric pollution;
(*g*) Countries with areas with fragile ecosystems, including mountainous ecosystems;
(*h*) Countries whose economies are highly dependent on income generated from the production, processing and export, and/or on consumption of fossil fuels and associated energy-intensive products; and
(*i*) Land-locked and transit countries.

Further, the Conference of the Parties may take actions, as appropriate, with respect to this paragraph.

(9) The Parties shall take full account of the specific needs and special situations of the least developed countries in their actions with regard to funding and transfer of technology.

(10) The Parties shall, in accordance with Article 10, take into consideration in the implementation of the commitments of the Convention the situation of Parties, particularly developing country Parties, with economies that are vulnerable to the adverse effects of the implementation of measures to respond to climate change. This applies notably to Parties with economies that are

highly dependent on income generated from the production, processing and export, and/or consumption of fossil fuels and associated energy-intensive products and/or the use of fossil fuels for which such Parties have serious difficulties in switching to alternatives.

Article 5. Research and Systematic Observation—In carrying out their commitments under Article 4, paragraph 1(*g*), the Parties shall:

(*a*) Support and further develop, as appropriate, international and intergovernmental programmes and networks or organizations aimed at defining, conducting, assessing and financing research, data collection and systematic observation, taking into account the need to minimize duplication of effort;

(*b*) Support international and intergovernmental efforts to strengthen systematic observation and national scientific and technical research capacities and capabilities, particularly in developing countries, and to promote access to, and the exchange of, data and analyses thereof obtained from areas beyond national jurisdiction; and

(*c*) Take into account the particular concerns and needs of developing countries and cooperate in improving their endogenous capacities and capabilities to participate in the efforts referred to in subparagraphs (*a*) and (*b*) above.

Article 6. Education, Training and Public Awareness—In carrying out their commitments under Article 4, paragraph 1(*i*), the Parties shall:

(*a*) Promote and facilitate at the national and, as appropriate, subregional and regional levels, and in accordance with national laws and regulations, and within their respective capacities:

(*i*) The development and implementation of educational and public awareness programmes on climate change and its effects;

(*ii*) Public access to information on climate change and its effects;

(*iii*) Public participation in addressing climate change and its effects and developing adequate responses; and

(*iv*) Training of scientific, technical and managerial personnel.

(*b*) Cooperate in and promote, at the international level, and, where appropriate, using existing bodies:

(*i*) The development and exchange of educational and public awareness material on climate change and its effects; and

(*ii*) The development and implementation of education and training programmes, including the strengthening of national institutions and the exchange or secondment of personnel to train experts in this field, in particular for developing countries.

Article 7. Conference of the Parties—(1) A Conference of the Parties is hereby established.

(2) The Conference of the Parties, as the supreme body of this Convention, shall keep under regular review the implementation of the Convention and any related legal instruments that the Conference of the Parties may adopt, and shall make, within its mandate, the decisions necessary to promote the effective implementation of the Convention. To this end, it shall:

(*a*) Periodically examine the obligations of the Parties and the institutional arrangements under the Convention, in the light of the objective of the Convention, the experience gained in its implementation and the evolution of scientific and technological knowledge;

(*b*) Promote and facilitate the exchange of information on measures adopted by the Parties to address climate change and its effects, taking into account the differing circumstances, responsibilities and capabilities of the Parties and their respective commitments under the Convention;

(*c*) Facilitate, at the request of two or more Parties, the coordination of measures adopted by them to address climate change and its effects, taking into account the differing circumstances, responsibilities and capabilities of the Parties and their respective commitments under the Convention;

(*d*) Promote and guide, in accordance with the objective and provisions of the Convention, the development and periodic refinement of comparable methodologies, to be agreed on by the Conference of the Parties, *inter alia*, for preparing inventories of greenhouse gas emissions by sources and removals by sinks, and for evaluating the effectiveness of measures to limit the emissions and enhance the removals of these gases;

(*e*) Assess, on the basis of all information made available to it in accordance with the provisions of the Convention, the implementation of the Convention by the Parties, the overall effects of the measures taken pursuant to the Convention, in particular environmental, economic and social effects as well as their cumulative impacts and the extent to which progress towards the objective of the Convention is being achieved;

(*f*) Consider and adopt regular reports on the implementation of the Convention and ensure their publication;

(*g*) Make recommendations on any matters necessary for the implementation of the Convention;

(*h*) Seek to mobilize financial resources in accordance with Article 4, paragraphs 3, 4 and 5, and Article 11;

(*i*) Establish such subsidiary bodies as are deemed necessary for the implementation of the Convention; 19

(*j*) Review reports submitted by its subsidiary bodies and provide guidance to them;

(*k*) Agree upon and adopt, by consensus, rules of procedure and financial rules for itself and for any subsidiary bodies;

(*l*) Seek and utilize, where appropriate, the services and cooperation of, and information provided by, competent international organizations and intergovernmental and non-governmental bodies; and

(*m*) Exercise such other functions as are required for the achievement of the objective of the Convention as well as all other functions assigned to it under the Convention.

(3) The Conference of the Parties shall, at its first session, adopt its own rules of procedure as well as those of the subsidiary bodies established by the Convention, which shall include decision-making procedures for matters not already covered by decision-making procedures stipulated in the Convention. Such procedures may include specified majorities required for the adoption of particular decisions.

(4) The first session of the Conference of the Parties shall be convened by the interim secretariat referred to in Article 21 and shall take place not later than one year after the date of entry into force of the Convention. Thereafter, ordinary sessions of the

Conference of the Parties shall be held every year unless otherwise decided by the Conference of the Parties.

(5) Extraordinary sessions of the Conference of the Parties shall be held at such other times as may be deemed necessary by the Conference, or at the written request of any Party, provided that, within six months of the request being communicated to the Parties by the secretariat, it is supported by at least one third of the Parties.

(6) The United Nations, its specialized agencies and the International Atomic Energy Agency, as well as any State member thereof or observers thereto not Party to the Convention, may be represented at sessions of the Conference of the Parties as observers. Any body or agency, whether national or international, governmental or non-governmental, which is qualified in matters covered by the Convention, and which has informed the secretariat of its wish to be represented at a session of the Conference of the Parties as an observer, may be so admitted unless at least one third of the Parties present object. The admission and participation of observers shall be subject to the rules of procedure adopted by the Conference of the Parties.

Article 8. Secretariat—(1) A secretariat is hereby established.

(2) The functions of the secretariat shall be:

(*a*) To make arrangements for sessions of the Conference of the Parties and its subsidiary bodies established under the Convention and to provide them with services as required;

(*b*) To compile and transmit reports submitted to it;

(*c*) To facilitate assistance to the Parties, particularly developing country Parties, on request, in the compilation and communication of information required in accordance with the provisions of the Convention;

(*d*) To prepare reports on its activities and present them to the Conference of the Parties;

(*e*) To ensure the necessary coordination with the secretariats of other relevant international bodies;

(*f*) To enter, under the overall guidance of the Conference of the Parties, into such administrative and contractual arrangements as may be required for the effective discharge of its functions; and

(*g*) To perform the other secretariat functions specified in the Convention and in any of its protocols and such other functions as may be determined by the Conference of the Parties.

(3) The Conference of the Parties, at its first session, shall designate a permanent secretariat and make arrangements for its functioning.

Article 9. Subsidiary Body for Scientific and Technological Advice—(1) A subsidiary body for scientific and technological advice is hereby established to provide the Conference of the Parties and, as appropriate, its other subsidiary bodies with timely information and advice on scientific and technological matters relating to the Convention. This body shall be open to participation by all Parties and shall be multidisciplinary. It shall comprise government representatives competent in the relevant field of expertise. It shall report regularly to the Conference of the Parties on all aspects of its work.

(2) Under the guidance of the Conference of the Parties, and drawing upon existing competent international bodies, this body shall:

(*a*) Provide assessments of the state of scientific knowledge relating to climate change and its effects;

(*b*) Prepare scientific assessments on the effects of measures taken in the implementation of the Convention;

(*c*) Identify innovative, efficient and state-of-the-art technologies and know-how and advise on the ways and means of promoting development and/or transferring such technologies;

(*d*) Provide advice on scientific programmes, international cooperation in research and development related to climate change, as well as on ways and means of supporting endogenous capacity building in developing countries; and

(*e*) Respond to scientific, technological and methodological questions that the Conference of the Parties and its subsidiary bodies may put to the body.

(3) The functions and terms of reference of this body may be further elaborated by the Conference of the Parties.

Article 10. Subsidiary Body for Implementation—(1) A subsidiary body for implementation is hereby established to assist the Conference of the Parties in the assessment and review of the effective implementation of the Convention. This body shall be open to participation by all Parties and comprise government representatives who are experts on matters related to climate change. It shall report regularly to the Conference of the Parties on all aspects of its work.

(2) Under the guidance of the Conference of the Parties, this body shall:

(*a*) Consider the information communicated in accordance with Article 12, paragraph 1, to assess the overall aggregated effect of the steps taken by the Parties in the light of the latest scientific assessments concerning climate change;

(*b*) Consider the information communicated in accordance with Article 12, paragraph 2, in order to assist the Conference of the Parties in carrying out the reviews required by Article 4, paragraph 2(*d*); and

(*c*) Assist the Conference of the Parties, as appropriate, in the preparation and implementation of its decisions.

Article 11. Financial Mechanism—(1) A mechanism for the provision of financial resources on a grant or concessional basis, including for the transfer of technology, is hereby defined. It shall function under the guidance of and be accountable to the Conference of the Parties, which shall decide on its policies, programme priorities and eligibility criteria related to this Convention. Its operation shall be entrusted to one or more existing international entities.

(2) The financial mechanism shall have an equitable and balanced representation of all Parties within a transparent system of governance.

(3) The Conference of the Parties and the entity or entities entrusted with the operation of the financial mechanism shall agree upon arrangements to give effect to the above paragraphs, which shall include the following:

(*a*) Modalities to ensure that the funded projects to address climate change are in conformity with the policies, programme priorities and eligibility criteria established by the Conference of the Parties;

(*b*) Modalities by which a particular funding decision may be reconsidered in light of these policies, programme priorities and eligibility criteria;

(*c*) Provision by the entity or entities of regular reports to the Conference of the Parties on its funding operations, which is consistent with the requirement for accountability set out in paragraph 1 above; and

(*d*) Determination in a predictable and identifiable manner of the amount of funding necessary and available for the implementation of this Convention and the conditions under which that amount shall be periodically reviewed.

(4) The Conference of the Parties shall make arrangements to implement the above-mentioned provisions at its first session, reviewing and taking into account the interim arrangements referred to in Article 21, paragraph 3, and shall decide whether these interim arrangements shall be maintained. Within four years thereafter, the Conference of the Parties shall review the financial mechanism and take appropriate measures.

(5) The developed country Parties may also provide and developing country Parties avail themselves of, financial resources related to the implementation of the Convention through bilateral, regional and other multilateral channels.

Article 12. Communication of Information Related to Implementation—(1) In accordance with Article 4, paragraph 1, each Party shall communicate to the Conference of the Parties, through the secretariat, the following elements of information:

(*a*) A national inventory of anthropogenic emissions by sources and removals by sinks of all greenhouse gases not controlled by the Montreal Protocol, to the extent its capacities permit, using comparable methodologies to be promoted and agreed upon by the Conference of the Parties;

(*b*) A general description of steps taken or envisaged by the Party to implement the Convention; and

(*c*) Any other information that the Party considers relevant to the achievement of the objective of the Convention and suitable for inclusion in its communication, including, if feasible, material relevant for calculations of global emission trends.

(2) Each developed country Party and each other Party included in Annex I shall incorporate in its communication the following elements of information:

(*a*) A detailed description of the policies and measures that it has adopted to implement its commitment under Article 4, paragraphs 2(*a*) and 2(*b*); and (*b*) A specific estimate of the effects that the policies and measures referred to in subparagraph (*a*) immediately above will have on anthropogenic emissions by its sources and removals by its sinks of greenhouse gases during the period referred to in Article 4, paragraph 2(*a*).

(3) In addition, each developed country Party and each other developed Party included in Annex II shall incorporate details of measures taken in accordance with Article 4, paragraphs 3, 4 and 5.

(4) Developing country Parties may, on a voluntary basis, propose projects for financing, including specific technologies, materials, equipment, techniques or practices that would be needed to implement such projects, along with, if possible, an estimate of all incremental costs, of the reductions of emissions and increments of removals of greenhouse gases, as well as an estimate of the consequent benefits.

(5) Each developed country Party and each other Party included in Annex I shall make its initial communication within six months of the entry into force of the Convention for that Party. Each Party not so listed shall make its initial communication within three years of the entry into force of the Convention for that Party, or of the availability of financial resources in accordance with Article 4, paragraph 3. Parties that are least developed countries may make their initial communication at their discretion. The frequency of subsequent communications by all Parties shall be determined by the Conference of the Parties, taking into account the differentiated timetable set by this paragraph.

(6) Information communicated by Parties under this Article shall be transmitted by the secretariat as soon as possible to the Conference of the Parties and to any subsidiary bodies concerned. If necessary, the procedures for the communication of information may be further considered by the Conference of the Parties.

(7) From its first session, the Conference of the Parties shall arrange for the provision to developing country Parties of technical and financial support, on request, in compiling and communicating information under this Article, as well as in identifying the technical and financial needs associated with proposed projects and response measures under Article 4. Such support may be provided by other Parties, by competent international organizations and by the secretariat, as appropriate.

(8) Any group of Parties may, subject to guidelines adopted by the Conference of the Parties, and to prior notification to the Conference of the Parties, make a joint communication in fulfilment of their obligations under this Article, provided that such a communication includes information on the fulfilment by each of these Parties of its individual obligations under the Convention.

(9) Information received by the secretariat that is designated by a Party as confidential, in accordance with criteria to be established by the Conference of the Parties, shall be aggregated by the secretariat to protect its confidentiality before being made available to any of the bodies involved in the communication and review of information.

(10) Subject to paragraph 9 above, and without prejudice to the ability of any Party to make public its communication at any time, the secretariat shall make communications by Parties under this Article publicly available at the time they are submitted to the Conference of the Parties.

Article 13. Resolution of Questions Regarding Implementation—The Conference of the Parties shall, at its first session, consider the establishment of a multilateral consultative process, available to Parties on their request, for the resolution of questions regarding the implementation of the Convention.

Article 14. Settlement of Disputes—(1) In the event of a dispute between any two or more Parties concerning the interpretation or application of the Convention, the Parties concerned shall seek a settlement of the dispute through negotiation or any other peaceful means of their own choice.

(2) When ratifying, accepting, approving or acceding to the Convention, or at any time thereafter, a Party which is not a regional economic integration organization may declare in a written instrument submitted to the Depositary that, in respect of any dispute concerning the interpretation or application of the Convention, it recognizes as compulsory ipso facto and without special agreement, in relation to any Party accepting the same obligation:

(*a*) Submission of the dispute to the International Court of Justice, and/or

(*b*) Arbitration in accordance with procedures to be adopted by the Conference of the Parties as soon as practicable, in an annex on arbitration.

A Party which is a regional economic integration organization may make a declaration with like effect in relation to arbitration in accordance with the procedures referred to in subparagraph (*b*) above.

(3) A declaration made under paragraph 2 above shall remain in force until it expires in accordance with its terms or until three months after written notice of its revocation has been deposited with the Depositary.

(4) A new declaration, a notice of revocation or the expiry of a declaration shall not in any way affect proceedings pending before the International Court of Justice or the arbitral tribunal, unless the parties to the dispute otherwise agree.

(5) Subject to the operation of paragraph 2 above, if after twelve months following notification by one Party to another that a dispute exists between them, the Parties concerned have not been able to settle their dispute through the means mentioned in paragraph 1 above, the dispute shall be submitted, at the request of any of the parties to the dispute, to conciliation.

(6) A conciliation commission shall be created upon the request of one of the parties to the dispute. The commission shall be composed of an equal number of members appointed by each party concerned and a chairman chosen jointly by the members appointed by each party. The commission shall render a recommendatory award, which the parties shall consider in good faith.

(7) Additional procedures relating to conciliation shall be adopted by the Conference of the Parties, as soon as practicable, in an annex on conciliation.

(8) The provisions of this Article shall apply to any related legal instrument which the Conference of the Parties may adopt, unless the instrument provides otherwise.

Article 15. Amendments to the Convention—(1) Any Party may propose amendments to the Convention.

(2) Amendments to the Convention shall be adopted at an ordinary session of the Conference of the Parties. The text of any proposed amendment to the Convention shall be communicated to the Parties by the secretariat at least six months before the meeting at which it is proposed for adoption. The secretariat shall also communicate proposed amendments to the signatories to the Convention and, for information, to the Depositary.

(3) The Parties shall make every effort to reach agreement on any proposed amendment to the Convention by consensus. If all efforts at consensus have been exhausted, and no agreement reached, the amendment shall as a last resort be adopted by a three-fourths majority vote of the Parties present and voting at the meeting. The adopted amendment shall be communicated by the secretariat to the Depositary, who shall circulate it to all Parties for their acceptance.

(4) Instruments of acceptance in respect of an amendment shall be deposited with the Depositary. An amendment adopted in accordance with paragraph 3 above shall enter into force for those Parties having accepted it on the ninetieth day after the date of receipt by the Depositary of an instrument of acceptance by at least three fourths of the Parties to the Convention.

(5) The amendment shall enter into force for any other Party on the ninetieth day after the date on which that Party deposits with the Depositary its instrument of acceptance of the said amendment.

(6) For the purposes of this Article, "Parties present and voting" means Parties present and casting an affirmative or negative vote.

Article 16. Adoption and Amendment of Annexes to the Convention—(1) Annexes to the Convention shall form an integral part thereof and, unless otherwise expressly provided, a reference to the Convention constitutes at the same time a reference to any annexes thereto. Without prejudice to the provisions of Article 14, paragraphs 2(*b*) and 7, such annexes shall be restricted to lists, forms and any other material of a descriptive nature that is of a scientific, technical, procedural or administrative character.

(2) Annexes to the Convention shall be proposed and adopted in accordance with the procedure set forth in Article 15, paragraphs 2, 3 and 4.

(3) An annex that has been adopted in accordance with paragraph 2 above shall enter into force for all Parties to the Convention six months after the date of the communication by the Depositary to such Parties of the adoption of the annex, except for those Parties that have notified the Depositary, in writing, within that period of their non-acceptance of the annex. The annex shall enter into force for Parties which withdraw their notification of non-acceptance on the ninetieth day after the date on which withdrawal of such notification has been received by the Depositary.

(4) The proposal, adoption and entry into force of amendments to annexes to the Convention shall be subject to the same procedure as that for the proposal, adoption and entry into force of annexes to the Convention in accordance with paragraphs 2 and 3 above.

(5) If the adoption of an annex or an amendment to an annex involves an amendment to the Convention, that annex or amendment to an annex shall not enter into force until such time as the amendment to the Convention enters into force.

Article 17. Protocols—(1) The Conference of the Parties may, at any ordinary session, adopt protocols to the Convention.

(2) The text of any proposed protocol shall be communicated to the Parties by the secretariat at least six months before such a session.

(3) The requirements for the entry into force of any protocol shall be established by that instrument.

(4) Only Parties to the Convention may be Parties to a protocol.

(5) Decisions under any protocol shall be taken only by the Parties to the protocol concerned.

Article 18. Right to Vote—(1) Each Party to the Convention shall have one vote, except as provided for in paragraph 2 below.

(2) Regional economic integration organizations, in matters within their competence, shall exercise their right to vote with a number of votes equal to the number of their member States that are Parties to the Convention. Such an organization shall not exercise its right to vote if any of its member States exercises its right, and vice versa.

Article 19. Depositary—The Secretary-General of the United Nations shall be the Depositary of the Convention and of protocols adopted in accordance with Article 17.

Article 20. Signature—This Convention shall be open for signature by States Members of the United Nations or of any of its specialized agencies or that are Parties to the Statute of the International Court of Justice and by regional economic integration organizations at Rio de Janeiro, during the United Nations Conference on Environment and Development, and thereafter at United Nations Headquarters in New York from 20 June 1992 to 19 June 1993.

Article 21. Interim Arrangements—(1) The secretariat functions referred to in Article 8 will be carried out on an interim basis by the secretariat established by the General Assembly of the

United Nations in its resolution 45/212 of 21 December 1990, until the completion of the first session of the Conference of the Parties.

(2) The head of the interim secretariat referred to in paragraph 1 above will cooperate closely with the Intergovernmental Panel on Climate Change to ensure that the Panel can respond to the need for objective scientific and technical advice. Other relevant scientific bodies could also be consulted.

(3) The Global Environment Facility of the United Nations Development Programme, the United Nations Environment Programme and the International Bank for Reconstruction and Development shall be the international entity entrusted with the operation of the financial mechanism referred to in Article 11 on an interim basis. In this connection, the Global Environment Facility should be appropriately restructured and its membership made universal to enable it to fulfil the requirements of Article 11.

Article 22. Ratification, Acceptance, Approval or Accession—(1) The Convention shall be subject to ratification, acceptance, approval or accession by States and by regional economic integration organizations. It shall be open for accession from the day after the date on which the Convention is closed for signature. Instruments of ratification, acceptance, approval or accession shall be deposited with the Depositary.

(2) Any regional economic integration organization which becomes a Party to the Convention without any of its member States being a Party shall be bound by all the obligations under the Convention. In the case of such organizations, one or more of whose member States is a Party to the Convention, the organization and its member States shall decide on their respective responsibilities for the performance of their obligations under the Convention. In such cases, the organization and the member States shall not be entitled to exercise rights under the Convention concurrently.

(3) In their instruments of ratification, acceptance, approval or accession, regional economic integration organizations shall declare the extent of their competence with respect to the matters governed by the Convention. These organizations shall also inform the Depositary, who shall in turn inform the Parties, of any substantial modification in the extent of their competence.

Article 23. Entry Into Force—(1) The Convention shall enter into force on the ninetieth day after the date of deposit of the fiftieth instrument of ratification, acceptance, approval or accession.

(2) For each State or regional economic integration organization that ratifies, accepts or approves the Convention or accedes thereto after the deposit of the fiftieth instrument of ratification, acceptance, approval or accession, the Convention shall enter into force on the ninetieth day after the date of deposit by such State or regional economic integration organization of its instrument of ratification, acceptance, approval or accession.

(3) For the purposes of paragraphs 1 and 2 above, any instrument deposited by a regional economic integration organization shall not be counted as additional to those deposited by States members of the organization.

Article 24. Reservations—No reservations may be made to the Convention.

Article 25. Withdrawal—(1) At any time after three years from the date on which the Convention has entered into force for a Party, that Party may withdraw from the Convention by giving written notification to the Depositary.

(2) Any such withdrawal shall take effect upon expiry of one year from the date of receipt by the Depositary of the notification of withdrawal, or on such later date as may be specified in the notification of withdrawal.

(3) Any Party that withdraws from the Convention shall be considered as also having withdrawn from any protocol to which it is a Party.

Article 26. Authentic Texts—(1) The original of this Convention, of which the Arabic, Chinese, English, French, Russian and Spanish texts are equally authentic, shall be deposited with the Secretary-General of the United Nations.

In Witness Whereof the undersigned, being duly authorized to that effect, have signed this Convention.

Done at New York this ninth day of May one thousand nine hundred and ninety-two.

Annex I

Australia
Austria
Belarus
Belgium
Bulgaria
Canada
Czechoslovakia
Denmark
European Economic Community
Estonia

Finland
France
Germany
Greece
Hungary
Iceland
Ireland
Italy
Japan
Latvia
Lithuania
Luxembourg
Netherlands
New Zealand
Norway
Poland
Portugal
Romania
Russian Federation
Spain
Sweden
Switzerland
Turkey
Ukraine
United Kingdom of Great Britain and Northern Ireland
United States of America

Annex II

Australia
Austria
Belgium
Canada
Denmark
European Economic Community
Finland
France
Germany
Greece
Iceland
Ireland
Italy
Japan
Luxembourg
Netherlands
New Zealand
Norway
Portugal
Spain
Sweden
Switzerland
Turkey
United Kingdom of Great Britain and Northern Ireland

Facing and Surveying the Problem

A major accomplishment of the Convention, which is general and flexible in character, is that it recognizes that there is a problem. That was no small thing a decade ago, when the treaty took effect and less scientific evidence was available. (And there are still those who dispute that global warming is real and that climate change is a problem.) It is hard to get the nations of the world to agree on anything, let alone a common approach to a difficulty which is complicated, whose consequences aren't entirely clear, and which will have its most severe effects decades and even centuries in the future.

The Convention sets an ultimate objective of stabilizing greenhouse gas emissions "at a level that would prevent dangerous anthropogenic (human induced) interference with the climate system." It states that "such a level should be achieved within a time-frame sufficient to allow ecosystems to adapt naturally to climate change, to ensure that food production is not threatened, and to enable economic development to proceed in a sustainable manner."

The Convention requires precise and regularly updated inventories of greenhouse gas emissions from industrialized countries. The first step in solving a problem is knowing its dimensions. With a few exceptions, the "base year" for tabulating greenhouse gas emissions has been set as 1990. Developing countries also are encouraged to carry out inventories.

Countries ratifying the treaty—called "Parties to the Convention" in diplomatic jargon—agree to take climate change into account in such matters as agriculture, industry, energy, natural resources, and activities involving sea coasts. They agree to develop national programmes to slow climate change.

The Convention recognizes that it is a "framework" document—something to be amended or augmented over time so that efforts to deal with global warming and climate change can be focused and made more effective. The first addition to the treaty, the Kyoto Protocol, was adopted in 1997.

Responsibility and Vulnerability

The Convention places the heaviest burden for fighting climate change on industrialized nations, since they are the source of most past and current greenhouse gas emissions. These countries are asked to do the most to cut what comes out of smokestacks and tailpipes, and to provide most of the money for efforts elsewhere. For the most part, these developed nations, called "Annex I" countries because they are listed in the first annex to the treaty, belong to the Organization for Economic Cooperation and Development (OECD).

These advanced nations, as well as 12 "economies in transition" (countries in Central and Eastern Europe, including some states formerly belonging to the Soviet Union) were expected by the year 2000 to reduce emissions to 1990 levels. As a group, they succeeded.

Industrialized nations agree under the Convention to support climate-change activities in developing countries by providing financial support above and beyond any financial assistance they already provide to these countries. A system of grants and loans has been set up through the Convention and is managed by the Global Environment Facility. Industrialized countries also agree to share technology with less-advanced nations.

Because economic development is vital for the world's poorer countries—and because such progress is difficult to achieve even without the complications added by climate change—the Convention accepts that the share of greenhouse gas emissions produced by developing nations will grow in the coming years. It nonetheless seeks to help such countries limit emissions in ways that will not hinder their economic progress.

The Convention acknowledges the vulnerability of developing countries to climate change and calls for special efforts to ease the consequences.

The Clean Development Mechanism: Getting Developing Countries Involved

The Kyoto Protocol does not set limits on the greenhouse-gas emissions of developing nations. Yet the greenhouse-gas emissions of developing countries are growing, especially in the case of enormously populous states such as China and India, which are rapidly expanding their industrial output.

Because the atmosphere is equally damaged by greenhouse-gas emissions wherever they occur and equally helped by emissions cuts wherever they are made, the Protocol includes an arrangement for reductions to be "sponsored" in countries not bound by emissions targets. The so-called Clean Development Mechanism is loaded with complicated details and acroynyms, but in simplified form it works this way: Industrialized countries pay for projects that cut or avoid emissions in poorer nations—and are awarded credits that can be applied to meeting their own emissions targets. The recipient countries benefit from free infusions of advanced technology that allow their factories or electrical generating plants to operate more efficiently—and hence at lower costs and higher profits. And the atmosphere benefits because future emissions are lower than they would have been otherwise.

The mechanism has drawn extensive interest from rich and poor countries alike, and steps have been taken to put it into operation even before the Protocol takes effect. In particular, it is cost-effective and offers a degree of flexibility to industrialized countries trying to meet their targets. It can be more efficient for them to carry out environmentally useful work in developing countries than at home, where land, technology, and labor are generally more costly. The benefits to the climate are the same.

The system also appeals to private companies and investors. The mechanism is meant to work bottom-up—to proceed from individual proposals to approval by donor and recipient governments to the allocation of "certified emissions reduction" credits. Countries earning the credits may apply them to meeting their emissions limits, may "bank" them for use later, or may sell them to other industrialized countries under the Protocol's emissions-trading system. Private firms are interested in the mechanism because they may earn profits from proposing and carrying out such work and because they may develop good reputations for their technology which will lead to further sales. A possible benefit for everyone is that the potential for profits may lead these businesses to develop even more useful technologies.

The Clean Development Mechanism will be overseen by an Executive Board that has already been established and already has approved a series of "methodologies" for large-scale and small-scale projects. To be certified, a project must be approved by all involved parties, demonstrate a measurable and long-term ability to reduce emissions, and promise reductions that would be additional to any that would otherwise occur.

A special provision allows credits earned under clean-development schemes to be valid and "bankable" now, although the Protocol has yet to take legal effect.

Options to the programme are being considered. Less red tape, for example, may be required for small-scale projects, such as renewable energy facilities below 15 megawatts of installed capacity. Another proposal is to allow afforestation and reforestation projects to be included in the scheme.

Emissions Trading: The "Carbon Market"

The limits on greenhouse-gas emissions set by the Kyoto Protocol are a way of assigning monetary value to the earth's shared atmosphere—something that has been missing up to now. Nations that have contributed the most to global warming have tended to benefit directly in terms of greater business profits and higher standards of living, while they have not been held proportionately accountable for the damages caused by their emissions. The negative effects of climate change will be felt all over the world, and actually the consequences are expected to be most severe in least-developed nations which have produced few emissions.

The Kyoto Protocol sets limits on total emissions by the world's major economies, a prescribed number of "emission units." Individual industrialized countries will have mandatory emissions targets they must meet... but it is understood that some will do better than expected, coming in under their limits, while others will exceed them.

The Protocol allows countries that have emissions units to spare—emissions permitted them but not "used"—to sell this excess capacity to countries that are over their targets. This so-called "carbon market"—so-named because carbon dioxide is the most widely produced greenhouse gas, and because emissions of other greenhouse gases will be recorded and counted in terms of their "carbon dioxide equivalents"—is both flexible and realistic. Countries not meeting their commitments will be able to "buy" compliance... but the price may be steep. The higher the cost, the more pressure they will feel to use energy more efficiently and to research and promote the development of alternative sources of energy that have low or no emissions.

A global "stock market" where emissions units are bought and sold is simple in concept—but in practice the Protocol's emissions-trading system has been complicated to set up. The details, weren't specified in the Protocol, and so additional negotiations were held to hammer them out. These rules were among the workaday specifics included in the 2001 "Marrakesh Accords." The problems are clear: countries' actual emissions have to be monitored and guaranteed to be what they are reported to be; and precise records have to be kept of the trades carried out. Accordingly, "registries"—like bank accounts of a nation's emissions units—are being set up, along with "accounting procedures," an "international transactions log," and "expert review teams" to police compliance.

More than actual emissions units will be involved in trades and sales. Countries will get credit for reducing greenhouse-gas totals by planting or expanding forests ("removal units"); for carrying out "joint implementation projects" with other developed countries, usually countries with "transition economies"; and for projects under the Protocol's Clean Development Mechanism, which involves funding activities to reduce emissions by developing nations. Credits earned this way may be bought and sold in the emissions market or "banked" for future use.

Some national registry systems under the Protocol have already been set up, as countries are eager to "bank" emissions reductions already accomplished while they wait for the Protocol to win its final ratifications and become legally binding.

Smaller "carbon markets" are now being established by the European Union and other groups of countries and are expected to begin operating in advance of the Protocol's entry into force. These emissions-trading systems are intended to start the process and to link up with the Protocol's global market once it becomes operational.

Joint Implementation

Mutual Help for Countries with Emissions Targets

"Joint implementation" is a programme under the Kyoto Protocol that allows industrialized countries to meet part of their required cuts in greenhouse-gas emissions by paying for projects that reduce emissions in other industrialized countries. In practice, this will likely mean facilities built in the countries of Eastern Europe and the former Soviet Union—the "transition economies"—paid for by Western European and North American countries.

The sponsoring governments will receive credits that may be applied to their emissions targets; the recipient nations will gain foreign investment and advanced technology (but not credit toward meeting their own emissions caps; they have to do that themselves). The system has advantages of flexibility and efficiency. It often is cheaper to carry out energy-efficiency work in the transition countries, and to realize greater cuts in emissions by doing so. The atmosphere benefits wherever these reductions occur.

The operation of the joint implementation mechanism is similar to that of the "clean development mechanism"—and similarly complicated. To go ahead with joint implementation projects, industralized countries must meet requirements under the Protocol for accurate inventories of greenhouse-gas emissions and for detailed registries of emissions "units" and "credits" (steps that also are required for the international trading of emissions on the "carbon market"). If these requirements are met, countries may carry out projects and receive credits beginning in 2008.

A pilot phase begun in 1995 allowed countries to gain experience in cooperating and in sharing technology. Most of the numerous pilot projects carried out will not be translated into credits under the Protocol, but schemes begun after 1 January 2002 which meet all requirements may be registered under the joint implementation programme.

If industrialized countries have not yet set up approved registries and greenhouse-gas inventory systems—complicated technical and bureaucratic hoops some nations still have to jump through—they may carry out joint-implementation projects under a "second track" process that involves greater international oversight. The oversight, which may be assigned to private companies, will ensure that emissions actually are reduced, and will certify by how much.

The Kyoto Protocol

This international agreement, which builds on the United Nations Framework Convention on Climate Change, sets legally binding targets and timetables for cutting the greenhouse-gas emissions of industrialized countries.

A summary of the Protocol. The Protocol, with its stiffer requirements, will have a real impact on climate change.

Emissions trading. Under the Protocol, countries may buy and sell greenhouse-gas emissions "units" and "credits."

Clean development mechanism. The Protocol provides a system for financing emissions-reducing or emissions-avoiding projects in developing nations.

Joint implementation. Within the Protocol, industrialized countries are granted "emissions reduction units" for financing projects in other developed countries—a system likely to increase efficiency and reduce the global-warming output of the "transition economies" of central and eastern Europe.

A Summary of the Kyoto Protocol: Powerful, Complicated... and Taking Root

It took all of one year for the member countries of the Framework Convention on Climate Change to decide that the Convention had to be augmented by an agreement with stricter demands for reducing greenhouse-gas emissions. The Convention took effect in 1994, and by 1995 governments had begun negotiations on a protocol—an international agreement linked to the existing treaty, but standing on its own. The text of the Kyoto Protocol was adopted unanimously in 1997; it entered into force on 16 February 2005.

The Protocol's major feature is that it has mandatory targets on greenhouse-gas emissions for the world's leading economies which have accepted it. These targets range from 8 per cent to +10 per cent of the countries' individual 1990 emissions levels "with a view to reducing their overall emissions of such gases by at least 5 per cent below existing 1990 levels in the commitment period 2008 to 2012." In almost all cases—even those set at +10 per cent of 1990 levels—the limits call for significant reductions in currently projected emissions. Future mandatory targets are expected to be established for "commitment periods" after 2012. These are to be negotiated well in advance of the periods concerned.

Commitments under the Protocol vary from nation to nation. The overall 5 per cent target for developed countries is to be met through cuts (from 1990 levels) of 8 per cent in the European Union [EU(15)], Switzerland, and most Central and East European states; 6 per cent in Canada; 7 per cent in the United States (although the US has since withdrawn its support for the Protocol); and 6 per cent in Hungary, Japan, and Poland. New Zealand, Russia, and Ukraine are to stabilize their emissions, while Norway may increase emissions by up to 1 per cent, Australia by up to 8 per cent (subsequently withdrew its support for the Protocol), and Iceland by 10 per cent. The EU has made its own internal agrement to meet its 8 per cent target by distributing different rates to its member states. These targets range from a 28 per cent reduction by Luxembourg and 21 per cent cuts by Denmark and Germany to a 25 per cent increase by Greece and a 27 per cent increase by Portugal.

To compensate for the sting of "binding targets," as they are called, the agreement offers flexibility in how countries may meet their targets. For example, they may partially compensate for their emissions by increasing "sinks"—forests, which remove carbon dioxide from the atmosphere. That may be accomplished either on their own territories or in other countries. Or they may pay for foreign projects that result in greenhouse-gas cuts. Several mechanisms have been set up for this purpose.

The Kyoto Protocol is a complicated agreement that has been slow in coming—there are reasons for this. The Protocol not only has to be an effective against a complicated worldwide problem—it also has to be politically acceptable. As a result, panels and committees have multiplied to monitor and referee its various programmes, and even after the agreement was approved in 1997, further negotiations were deemed necessary to hammer out instructions on how to "operate" it. These rules, adopted in 2001, are called the "Marrakesh Accords."

There is a delicate balance to international treaties. Those appealing enough to gain widespread support often aren't strong enough to solve the problems they focus on. (Because the Framework Convention was judged to have this weakness, despite its many valuable provisions, the Protocol was created to supplement it.) Yet treaties with real "teeth" may have difficulty attracting enough widespread support to be effective.

A positive development, meanwhile, is that some mechanisms of the Protocol have enough support that they are being set up in advance of the Protocol's entry into force. The Clean Development Mechanism, for example—through which industrialized countries can partly meet their binding emissions targets through "credits" earned by sponsoring greenhouse-gas-reducing projects in developing countries—already has an executive board that is sifting through proposals.

Bodies of the Framework Convention, Actors in the Negotiation Process, and the UNFCCC Secretariat

Bodies of the Convention and Partner Agencies

The Conference of the Parties (COP) is the prime authority of the Convention. It is an association of all member countries (or

"Parties") and usually meets annually for a period of two weeks. These sessions are attended by several thousand government delegates, observers, and journalists. The Conference of the Parties evaluates the status of climate change and the effectiveness of the treaty. It examines the activities of member countries, particularly by reviewing national communications and emissions inventories; it considers new scientific findings; and it tries to capitalize on experience as efforts to address climate change proceed.

A Subsidiary Body for Scientific and Technological Advice (SBSTA) counsels the Conference of the Parties on matters of climate, the environment, technology, and method. It meets twice a year.

A Subsidiary Body for Implementation (SBI) helps review how the Convention is being applied, for example by analyzing the national communications submitted by member countries. It also deals with financial and administrative matters. The SBI meets twice each year.

Several expert groups exist under the Convention. A Consultative Group of Experts on National Communications from "Non-Annex 1 Parties" helps developing countries prepare national reports on climate change issues. A Least Developed Country Expert Group advises such nations on establishing programmes for adapting to climate change. And an Expert Group on Technology Transfer seeks to spur the sharing of technology with less-advanced nations.

Partner agencies include the Global Environment Facility (GEF), which has existed since 1991 to fund projects in developing countries that will have global environmental benefits. The job of channeling grants and loans to poor countries to help them address climate change, as called for by the Convention, has been delegated to the GEF because of its established expertise. And the Intergovernmental Panel on Climate Change (IPCC) provides services to the Convention, although it is not part of it, through publishing comprehensive reviews every five years of the status of climate change and climate-change science, along with special reports and technical papers on request.

Actors in the Negotiation Process

Countries belonging to the Convention hold the real power—they take decisions at sessions of the Conference of the Parties (most decisions are reached by consensus). Member countries often form alliances to increase efficiency and maximize influence during negotiations. The Conference has several groupings representing the concerns of developing countries, least-developed countries, small-island states, Europe (through the European Union), non-European industrialized nations, oil-exporting nations, and nations committed to "environmental integrity."

Countries get extensive input from other sources, both through official channels and in behind-the-scenes chatter. This is not surprising, considering that the global climate is facing a major threat, coastlines and even countries may disappear, and industries and livelihoods may wax or wane... not to mention that millions of dollars are being allocated for programmes and activities.

"Observer" is the official—and misleadlingly quiet-sounding—term for groups and agencies allowed to attend and even speak at international meetings, but not to participate in decision-making. Among observers permitted by the Convention are intergovernmental agencies, such as the United Nations Development Programme (UNDP), the United Nations Environment Programme (UNEP), the World Meteorological Organization (WMO); the Organization for Economic Cooperation and Development (OECD); the International Energy Agency; and the Organization of Petroleum Exporting Countries (OPEC). To date, over 50 intergovernmental agencies and international organizations attend sessions of the Conference of Parties.

Observers also include a lively crowd of non-governmental organizations, known as "NGOs." These represent business and industrial interests, environmental groups, local governments, research and academic institutes, religious bodies, labour organizations, and population groups such as indigenous peoples. To win accreditation as observers, NGOs must be legally constituted not-for-profit entities "competent in matters related to the Convention." Currently, more than 600 NGOs are accredited to participate in meetings related to the Convention.

The UNFCCC Secretariat

A secretariat staffed by international civil servants supports the Convention and its supporting bodies. It makes practical arrangements for meetings, compiles and distributes statistics and information, and assists member countries in meeting their commitments under the Convention. The secretariat is based in Bonn, Germany.

STATUS OF RATIFICATION

The text of the Convention was adopted at the United Nations Headquarters, New York on the 9 May 1992; it was open for signature at the Rio de Janeiro from 4 to 14 June 1992, and thereafter at the United Nations Headquarters, New York, from 20 June 1992 to 19 June 1993. By that date the Convention had received 166 signatures. The Convention entered into force on 21 March 1994. Those States that have not signed the Convention may accede to it at any time.

The list below contains the latest information concerning dates of signature and ratification received from the Secretary-General of the United Nations, as Depository of the Convention. The dates in the column entitled "date of ratification" are those of the receipt of the instrument of ratification (R), acceptance (At), approval (Ap) or accession (Ac).

List of Signaturies and Ratification of the Convention Parties (153 kb) in chronological order as of 24 May 2004. The Convention currently has received 189 instruments of ratification.

Chapter 2

Protecting the Forests and Forestry Around the World

REPORT OF THE UNITED NATIONS CONFERENCE ON ENVIRONMENT AND DEVELOPMENT

Non-Legally Binding Authoritative Statement of Principles for a Global Consensus on the Management, Conservation and Sustainable Development of All Types of Forests

Preamble

(*a*) The subject of forests is related to the entire range of environmental and development issues and opportunities, including the right to socio-economic development on a sustainable basis.

(*b*) The guiding objective of these principles is to contribute to the management, conservation and sustainable development of forests and to provide for their multiple and complementary functions and uses.

(*c*) Forestry issues and opportunities should be examined in a holistic and balanced manner within the overall context of environment and development, taking into consideration the multiple functions and uses of forests, including traditional uses, and the likely economic and social stress when these uses are constrained or restricted, as well as the potential for development that sustainable forest management can offer.

(*d*) These principles reflect a first global consensus on forests. In committing themselves to the prompt implementation of these principles, countries also decide to keep them under assessment for their adequacy with regard to further international cooperation on forest issues.

(*e*) These principles should apply to all types of forests, both natural and planted, in all geographical regions and climatic zones, including austral, boreal, subtemperate, temperate, subtropical and tropical.

(*f*) All types of forests embody complex and unique ecological processes which are the basis for their present and potential capacity to provide resources to satisfy human needs as well as environmental values, and as such their sound management and conservation is of concern to the Governments of the countries to which they belong and are of value to local communities and to the environment as a whole.

(*g*) Forests are essential to economic development and the maintenance of all forms of life.

(*h*) Recognizing that the responsibility for forest management, conservation and sustainable development is in many States allocated among federal/national, state/provincial and local levels of government, each State, in accordance with its constitution and/or national legislation, should pursue these principles at the appropriate level of government.

Principles/Elements

1. (*a*) States have, in accordance with the Charter of the United Nations and the principles of international law, the sovereign right to exploit their own resources pursuant to their own environmental policies and have the responsibility to ensure that activities within their jurisdiction or control do not cause damage to the environment of other States or of areas beyond the limits of national jurisdiction.

(*b*) The agreed full incremental cost of achieving benefits associated with forest conservation and sustainable development requires increased international cooperation and should be equitably shared by the international community.

2. (*a*) States have the sovereign and inalienable right to utilize, manage and develop their forests in accordance with their

development needs and level of socio-economic development and on the basis of national policies consistent with sustainable development and legislation, including the conversion of such areas for other uses within the overall socio-economic development plan and based on rational land-use policies.

(*b*) Forest resources and forest lands should be sustainably managed to meet the social, economic, ecological, cultural and spiritual needs of present and future generations. These needs are for forest products and services, such as wood and wood products, water, food, fodder, medicine, fuel, shelter, employment, recreation, habitats for wildlife, landscape diversity, carbon sinks and reservoirs, and for other forest products. Appropriate measures should be taken to protect forests against harmful effects of pollution, including air-borne pollution, fires, pests and diseases, in order to maintain their full multiple value.

(*c*) The provision of timely, reliable and accurate information on forests and forest ecosystems is essential for public understanding and informed decision-making and should be ensured.

(*d*) Governments should promote and provide opportunities for the participation of interested parties, including local communities and indigenous people, industries, labour, non-governmental organizations and individuals, forest dwellers and women, in the development, implementation and planning of national forest policies.

3. (*a*) National policies and strategies should provide a framework for increased efforts, including the development and strengthening of institutions and programmes for the management, conservation and sustainable development of forests and forest lands.

(*b*) International institutional arrangements, building on those organizations and mechanisms already in existence, as appropriate, should facilitate international cooperation in the field of forests.

(*c*) All aspects of environmental protection and social and economic development as they relate to forests and forest lands should be integrated and comprehensive.

4. The vital role of all types of forests in maintaining the ecological processes and balance at the local, national, regional and global levels through, *inter alia*, their role in protecting fragile ecosystems, watersheds and freshwater resources and as rich storehouses of biodiversity and biological resources and sources of genetic material for biotechnology products, as well as photosynthesis, should be recognized.

5. (*a*) National forest policies should recognize and duly support the identity, culture and the rights of indigenous people, their communities and other communities and forest dwellers. Appropriate conditions should be promoted for these groups to enable them to have an economic stake in forest use, perform economic activities, and achieve and maintain cultural identity and social organization, as well as adequate levels of livelihood and well-being, through, *inter alia*, those land tenure arrangements which serve as incentives for the sustainable management of forests.

(*b*) The full participation of women in all aspects of the management, conservation and sustainable development of forests should be actively promoted.

6. (*a*) All types of forests play an important role in meeting energy requirements through the provision of a renewable source of bio-energy, particularly in developing countries, and the demands for fuelwood for household and industrial needs should be met through sustainable forest management, afforestation and reforestation. To this end, the potential contribution of plantations of both indigenous and introduced species for the provision of both fuel and industrial wood should be recognized.

(*b*) National policies and programmes should take into account the relationship, where it exists, between the conservation, management and sustainable development of forests and all aspects related to the production, consumption, recycling and/or final disposal of forest products.

(*c*) Decisions taken on the management, conservation and sustainable development of forest resources should benefit, to the extent practicable, from a comprehensive assessment of economic and non-economic values of forest goods and services and of the environmental costs and benefits. The development and improvement of methodologies for such evaluations should be promoted.

(*d*) The role of planted forests and permanent agricultural crops as sustainable and environmentally sound sources of renewable energy and industrial raw material should be recognized, enhanced and promoted. Their contribution to the maintenance of ecological processes, to offsetting pressure on primary/old-growth forest and to providing regional employment and development with the adequate involvement of local inhabitants should be recognized and enhanced.

(*e*) Natural forests also constitute a source of goods and services, and their conservation, sustainable management and use should be promoted.

7. (*a*) Efforts should be made to promote a supportive international economic climate conducive to sustained and environmentally sound development of forests in all countries, which include, *inter alia*, the promotion of sustainable patterns of production and consumption, the eradication of poverty and the promotion of food security.

(*b*) Specific financial resources should be provided to developing countries with significant forest areas which establish programmes for the conservation of forests including protected natural forest areas. These resources should be directed notably to economic sectors which would stimulate economic and social substitution activities.

8. (*a*) Efforts should be undertaken towards the greening of the world. All countries, notably developed countries, should take positive and transparent action towards reforestation, afforestation and forest conservation, as appropriate.

(*b*) Efforts to maintain and increase forest cover and forest productivity should be undertaken in ecologically, economically and socially sound ways through the rehabilitation, reforestation and re-establishment of trees and forests on unproductive, degraded and deforested lands, as well as through the management of existing forest resources.

(*c*) The implementation of national policies and programmes aimed at forest management, conservation and sustainable development, particularly in developing countries, should be supported by international financial and technical cooperation, including through the private sector, where appropriate.

(*d*) Sustainable forest management and use should be carried out in accordance with national development policies and priorities and on the basis of environmentally sound national guidelines. In the formulation of such guidelines, account should be taken, as appropriate and if applicable, of relevant internationally agreed methodologies and criteria.

(*e*) Forest management should be integrated with management of adjacent areas so as to maintain ecological balance and sustainable productivity.

(*f*) National policies and/or legislation aimed at management, conservation and sustainable development of forests should include the protection of ecologically viable representative or unique examples of forests, including primary/old-growth forests, cultural, spiritual, historical, religious and other unique and valued forests of national importance.

(*g*) Access to biological resources, including genetic material, shall be with due regard to the sovereign rights of the countries where the forests are located and to the sharing on mutually agreed terms of technology and profits from biotechnology products that are derived from these resources.

(*h*) National policies should ensure that environmental impact assessments should be carried out where actions are likely to have significant adverse impacts on important forest resources, and where such actions are subject to a decision of a competent national authority.

9. (*a*) The efforts of developing countries to strengthen the management, conservation and sustainable development of their forest resources should be supported by the international community, taking into account the importance of redressing external indebtedness, particularly where aggravated by the net transfer of resources to developed countries, as well as the problem of achieving at least the replacement value of forests through improved market access for forest products, especially processed products. In this respect, special

attention should also be given to the countries undergoing the process of transition to market economies.

(*b*) The problems that hinder efforts to attain the conservation and sustainable use of forest resources and that stem from the lack of alternative options available to local communities, in particular the urban poor and poor rural populations who are economically and socially dependent on forests and forest resources, should be addressed by Governments and the international community.

(*c*) National policy formulation with respect to all types of forests should take account of the pressures and demands imposed on forest ecosystems and resources from influencing factors outside the forest sector, and intersectoral means of dealing with these pressures and demands should be sought.

10. New and additional financial resources should be provided to developing countries to enable them to sustainably manage, conserve and develop their forest resources, including through afforestation, reforestation and combating deforestation and forest and land degradation.

11. In order to enable, in particular, developing countries to enhance their endogenous capacity and to better manage, conserve and develop their forest resources, the access to and transfer of environmentally sound technologies and corresponding know-how on favourable terms, including on concessional and preferential terms, as mutually agreed, in accordance with the relevant provisions of Agenda 21, should be promoted, facilitated and financed, as appropriate.

12. (*a*) Scientific research, forest inventories and assessments carried out by national institutions which take into account, where relevant, biological, physical, social and economic variables, as well as technological development and its application in the field of sustainable forest management, conservation and development, should be strengthened through effective modalities, including international cooperation. In this context, attention should also be given to research and development of sustainably harvested non-wood products.

(*b*) National and, where appropriate, regional and international institutional capabilities in education, training, science, technology, economics, anthropology and social aspects of forests and forest management are essential to the conservation and sustainable development of forests and should be strengthened.

(*c*) International exchange of information on the results of forest and forest management research and development should be enhanced and broadened, as appropriate, making full use of education and training institutions, including those in the private sector.

(*d*) Appropriate indigenous capacity and local knowledge regarding the conservation and sustainable development of forests should, through institutional and financial support and in collaboration with the people in the local communities concerned, be recognized, respected, recorded, developed and, as appropriate, introduced in the implementation of programmes. Benefits arising from the utilization of indigenous knowledge should therefore be equitably shared with such people.

13. (*a*) Trade in forest products should be based on non-discriminatory and multilaterally agreed rules and procedures consistent with international trade law and practices. In this context, open and free international trade in forest products should be facilitated.

(*b*) Reduction or removal of tariff barriers and impediments to the provision of better market access and better prices for higher value-added forest products and their local processing should be encouraged to enable producer countries to better conserve and manage their renewable forest resources.

(*c*) Incorporation of environmental costs and benefits into market forces and mechanisms, in order to achieve forest conservation and sustainable development, should be encouraged both domestically and internationally.

(*d*) Forest conservation and sustainable development policies should be integrated with economic, trade and other relevant policies.

(*e*) Fiscal, trade, industrial, transportation and other policies and practices that may lead to forest degradation should be avoided. Adequate policies, aimed at management, conservation and sustainable development of forests, including, where appropriate, incentives, should be encouraged.

14. Unilateral measures, incompatible with international obligations or agreements, to restrict and / or ban international trade in timber or other forest products should be removed or avoided, in order to attain long-term sustainable forest management.

15. Pollutants, particularly air-borne pollutants, including those responsible for acidic deposition, that are harmful to the health of forest ecosystems at the local, national, regional and global levels should be controlled.

Chapter 3

Global Guidelines on Biodiversity and Biosafety

CONVENTION ON BIOLOGICAL DIVERSITY

Preamble

The Contracting Parties,

Conscious of the intrinsic value of biological diversity and of the ecological, genetic, social, economic, scientific, educational, cultural, recreational and aesthetic values of biological diversity and its components,

Conscious also of the importance of biological diversity for evolution and for maintaining life sustaining systems of the biosphere,

Affirming that the conservation of biological diversity is a common concern of humankind,

Reaffirming that States have sovereign rights over their own biological resources,

Reaffirming also that States are responsible for conserving their biological diversity and for using their biological resources in a sustainable manner,

Concerned that biological diversity is being significantly reduced by certain human activities,

Aware of the general lack of information and knowledge regarding biological diversity and of the urgent need to develop scientific, technical and institutional capacities to provide the basic understanding upon which to plan and implement appropriate measures,

Noting that it is vital to anticipate, prevent and attack the causes of significant reduction or loss of biological diversity at source,

Noting also that where there is a threat of significant reduction or loss of biological diversity, lack of full scientific certainty should not be used as a reason for postponing measures to avoid or minimize such a threat,

Noting further that the fundamental requirement for the conservation of biological diversity is the *in-situ* conservation of ecosystems and natural habitats and the maintenance and recovery of viable populations of species in their natural surroundings,

Noting further that *ex-situ* measures, preferably in the country of origin, also have an important role to play,

Recognizing the close and traditional dependence of many indigenous and local communities embodying traditional lifestyles on biological resources, and the desirability of sharing equitably benefits arising from the use of traditional knowledge, innovations and practices relevant to the conservation of biological diversity and the sustainable use of its components,

Recognizing also the vital role that women play in the conservation and sustainable use of biological diversity and affirming the need for the full participation of women at all levels of policymaking and implementation for biological diversity conservation,

Stressing the importance of, and the need to promote, international, regional and global cooperation among States and intergovernmental organizations and the non-governmental sector for the conservation of biological diversity and the sustainable use of its components,

Acknowledging that the provision of new and additional financial resources and appropriate access to relevant technologies can be expected to make a substantial difference in the world's ability to address the loss of biological diversity,

Acknowledging further that special provision is required to meet the needs of developing countries, including the provision of new and additional financial resources and appropriate access to relevant technologies,

Noting in this regard the special conditions of the least developed countries and small island States,

Acknowledging that substantial investments are required to conserve biological diversity and that there is the expectation of a broad range of environmental, economic and social benefits from those investments,

Recognizing that economic and social development and poverty eradication are the first and overriding priorities of developing countries,

Aware that conservation and sustainable use of biological diversity is of critical importance for meeting the food, health and other needs of the growing world population, for which purpose access to and sharing of both genetic resources and technologies are essential,

Noting that, ultimately, the conservation and sustainable use of biological diversity will strengthen friendly relations among States and contribute to peace for humankind,

Desiring to enhance and complement existing international arrangements for the conservation of biological diversity and sustainable use of its components, and

Determined to conserve and sustainably use biological diversity for the benefit of present and future generations,

Have agreed as follows:

Article 1. Objectives—The objectives of this Convention, to be pursued in accordance with its relevant provisions, are the conservation of biological diversity, the sustainable use of its components and the fair and equitable sharing of the benefits arising out of the utilization of genetic resources, including by appropriate access to genetic resources and by appropriate transfer of relevant technologies, taking into account all rights over those resources and to technologies, and by appropriate funding.

Article 2. Use of Terms—For the purposes of this Convention:

"Biological diversity" means the variability among living organisms from all sources including, *inter alia*, terrestrial, marine and other aquatic ecosystems and the ecological complexes of which they are part; this includes diversity within species, between species and of ecosystems.

"Biological resources" includes genetic resources, organisms or parts thereof, populations, or any other biotic component of ecosystems with actual or potential use or value for humanity.

"Biotechnology" means any technological application that uses biological systems, living organisms, or derivatives thereof, to make or modify products or processes for specific use.

"Country of origin of genetic resources" means the country which possesses those genetic resources in *in-situ* conditions.

"*Country providing genetic resources*" means the country supplying genetic resources collected from *in-situ* sources, including populations of both wild and domesticated species, or taken from *ex-situ* sources, which may or may not have originated in that country.

"*Domesticated or cultivated species*" means species in which the evolutionary process has been influenced by humans to meet their needs.

"*Ecosystem*" means a dynamic complex of plant, animal and micro-organism communities and their non-living environment interacting as a functional unit.

"*Ex-situ conservation*" means the conservation of components of biological diversity outside their natural habitats.

"*Genetic material*" means any material of plant, animal, microbial or other origin containing functional units of heredity.

"*Genetic resources*" means genetic material of actual or potential value.

"*Habitat*" means the place or type of site where an organism or population naturally occurs.

"*In-situ conditions*" means conditions where genetic resources exist within ecosystems and natural habitats, and, in the case of domesticated or cultivated species, in the surroundings where they have developed their distinctive properties.

"*In-situ conservation*" means the conservation of ecosystems and natural habitats and the maintenance and recovery of viable populations of species in their natural surroundings and, in the case of domesticated or cultivated species, in the surroundings where they have developed their distinctive properties.

"*Protected area*" means a geographically defined area which is designated or regulated and managed to achieve specific conservation objectives.

"*Regional economic integration organization*" means an organization constituted by sovereign States of a given region, to which its member States have transferred competence in respect of matters governed by this Convention and which has been duly authorized, in accordance with its internal procedures, to sign, ratify, accept, approve or accede to it.

"*Sustainable use*" means the use of components of biological diversity in a way and at a rate that does not lead to the long-term decline of biological diversity, thereby maintaining its potential to meet the needs and aspirations of present and future generations.

"Technology" includes biotechnology.

Article 3. Principle—States have, in accordance with the Charter of the United Nations and the principles of international law, the sovereign right to exploit their own resources pursuant to their own environmental policies, and the responsibility to ensure that activities within their jurisdiction or control do not cause damage to the environment of other States or of areas beyond the limits of national jurisdiction.

Article 4. Jurisdictional Scope—Subject to the rights of other States, and except as otherwise expressly provided in this Convention, the provisions of this Convention apply, in relation to each Contracting Party:

(*a*) In the case of components of biological diversity, in areas within the limits of its national jurisdiction; and

(*b*) In the case of processes and activities, regardless of where their effects occur, carried out under its jurisdiction or control, within the area of its national jurisdiction or beyond the limits of national jurisdiction.

Article 5. Cooperation—Each Contracting Party shall, as far as possible and as appropriate, cooperate with other Contracting Parties, directly or, where appropriate, through competent international organizations, in respect of areas beyond national jurisdiction and on other matters of mutual interest, for the conservation and sustainable use of biological diversity.

Article 6. General Measures for Conservation and Sustainable Use—Each Contracting Party shall, in accordance with its particular conditions and capabilities:

(*a*) Develop national strategies, plans or programmes for the conservation and sustainable use of biological diversity or adapt for this purpose existing strategies, plans or programmes which shall reflect, *inter alia*, the measures set out in this Convention relevant to the Contracting Party concerned; and

(*b*) Integrate, as far as possible and as appropriate, the conservation and sustainable use of biological diversity into relevant sectoral or cross-sectoral plans, programmes and policies.

Article 7. Identification and Monitoring—Each Contracting Party shall, as far as possible and as appropriate, in particular for the purposes of Articles 8 to 10:

(*a*) Identify components of biological diversity important for its conservation and sustainable use having regard to the indicative list of categories set down in Annex I;

(*b*) Monitor, through sampling and other techniques, the components of biological diversity identified pursuant to subparagraph (*a*) above, paying particular attention to those requiring urgent conservation measures and those which offer the greatest potential for sustainable use;

(*c*) Identify processes and categories of activities which have or are likely to have significant adverse impacts on the conservation and sustainable use of biological diversity, and monitor their effects through sampling and other techniques; and

(*d*) Maintain and organize, by any mechanism data, derived from identification and monitoring activities pursuant to subparagraphs (*a*), (*b*) and (*c*) above.

Article 8. In-situ Conservation—Each Contracting Party shall, as far as possible and as appropriate:

(*a*) Establish a system of protected areas or areas where special measures need to be taken to conserve biological diversity;

(*b*) Develop, where necessary, guidelines for the selection, establishment and management of protected areas or areas where special measures need to be taken to conserve biological diversity;

(*c*) Regulate or manage biological resources important for the conservation of biological diversity whether within or outside protected areas, with a view to ensuring their conservation and sustainable use;

(*d*) Promote the protection of ecosystems, natural habitats and the maintenance of viable populations of species in natural surroundings;

(*e*) Promote environmentally sound and sustainable development in areas adjacent to protected areas with a view to furthering protection of these areas;

(*f*) Rehabilitate and restore degraded ecosystems and promote the recovery of threatened species, *inter alia*, through the development and implementation of plans or other management strategies;

(*g*) Establish or maintain means to regulate, manage or control the risks associated with the use and release of living modified organisms resulting from biotechnology which are likely to have adverse environmental impacts that could affect the conservation and sustainable use of biological diversity, taking also into account the risks to human health;

(*h*) Prevent the introduction of, control or eradicate those alien species which threaten ecosystems, habitats or species;

(*i*) Endeavour to provide the conditions needed for compatibility between present uses and the conservation of biological diversity and the sustainable use of its components;

(*j*) Subject to its national legislation, respect, preserve and maintain knowledge, innovations and practices of indigenous and local communities embodying traditional lifestyles relevant for the conservation and sustainable use of biological diversity and promote their wider application with the approval and involvement of the holders of such knowledge, innovations and practices and encourage the equitable sharing of the benefits arising from the utilization of such knowledge, innovations and practices;

(*k*) Develop or maintain necessary legislation and/or other regulatory provisions for the protection of threatened species and populations;

(*l*) Where a significant adverse effect on biological diversity has been determined pursuant to Article 7, regulate or manage the relevant processes and categories of activities; and

(*m*) Cooperate in providing financial and other support for *in situ* conservation outlined in subparagraphs (*a*) to (*l*) above, particularly to developing countries.

Article 9. Ex-situ Conservation—Each Contracting Party shall, as far as possible and as appropriate, and predominantly for the purpose of complementing *in-situ* measures:

(*a*) Adopt measures for the *ex-situ* conservation of components of biological diversity, preferably in the country of origin of such components;

(*b*) Establish and maintain facilities for *ex-situ* conservation of and research on plants, animals and micro-organisms, preferably in the country of origin of genetic resources;

(*c*) Adopt measures for the recovery and rehabilitation of threatened species and for their reintroduction into their natural habitats under appropriate conditions;

(*d*) Regulate and manage collection of biological resources from natural habitats for *ex-situ* conservation purposes so as not to threaten ecosystems and *in-situ* populations of species, except where special temporary *ex-situ* measures are required under subparagraph (*c*) above; and

(*e*) Cooperate in providing financial and other support for *ex situ* conservation outlined in subparagraphs (*a*) to (*d*) above and in the establishment and maintenance of *ex-situ* conservation facilities in developing countries.

Article 10. Sustainable Use of Components of Biological Diversity—Each Contracting Party shall, as far as possible and as appropriate:

(*a*) Integrate consideration of the conservation and sustainable use of biological resources into national decision-making;

(*b*) Adopt measures relating to the use of biological resources to avoid or minimize adverse impacts on biological diversity;

(*c*) Protect and encourage customary use of biological resources in accordance with traditional cultural practices that are compatible with conservation or sustainable use requirements;

(*d*) Support local populations to develop and implement remedial action in degraded areas where biological diversity has been reduced; and

(*e*) Encourage cooperation between its governmental authorities and its private sector in developing methods for sustainable use of biological resources.

Article 11. Incentive Measures—Each Contracting Party shall, as far as possible and as appropriate, adopt economically and socially sound measures that act as incentives for the conservation and sustainable use of components of biological diversity.

Article 12. Research and Training—The Contracting Parties, taking into account the special needs of developing countries, shall:

(*a*) Establish and maintain programmes for scientific and technical education and training in measures for the identification, conservation and sustainable use of biological diversity and its components and provide support for such education and training for the specific needs of developing countries;

(*b*) Promote and encourage research which contributes to the conservation and sustainable use of biological diversity, particularly in developing countries, *inter alia*, in accordance with decisions of the Conference of the Parties taken in consequence of recommendations of the Subsidiary Body on Scientific, Technical and Technological Advice; and

(*c*) In keeping with the provisions of Articles 16, 18 and 20, promote and cooperate in the use of scientific advances in biological diversity research in developing methods for conservation and sustainable use of biological resources.

Article 13. Public Education and Awareness—The Contracting Parties shall:

(*a*) Promote and encourage understanding of the importance of, and the measures required for, the conservation of biological diversity, as well as its propagation through media, and the inclusion of these topics in educational programmes; and

(*b*) Cooperate, as appropriate, with other States and international organizations in developing educational and public awareness programmes, with respect to conservation and sustainable use of biological diversity.

Article 14. Impact Assessment and Minimizing Adverse Impacts—(1) Each Contracting Party, as far as possible and as appropriate, shall:

(*a*) Introduce appropriate procedures requiring environmental impact assessment of its proposed projects that are likely to have significant adverse effects on biological diversity with a view to avoiding or minimizing such effects and, where appropriate, allow for public participation in such procedures;

(*b*) Introduce appropriate arrangements to ensure that the environmental consequences of its programmes and policies that are likely to have significant adverse impacts on biological diversity are duly taken into account;

(*c*) Promote, on the basis of reciprocity, notification, exchange of information and consultation on activities under their jurisdiction or control which are likely to significantly affect adversely the biological diversity of other States or areas beyond the limits of national jurisdiction, by encouraging the conclusion of bilateral, regional or multilateral arrangements, as appropriate;

(*d*) In the case of imminent or grave danger or damage, originating under its jurisdiction or control, to biological diversity within the area under jurisdiction of other States or in areas beyond the limits of national jurisdiction, notify immediately the potentially affected States of such danger or damage, as well as initiate action to prevent or minimize such danger or damage; and

(*e*) Promote national arrangements for emergency responses to activities or events, whether caused naturally or otherwise, which present a grave and imminent danger to biological diversity and encourage international cooperation to supplement such national efforts and, where appropriate and agreed by the States or regional economic integration organizations concerned, to establish joint contingency plans.

(2) The Conference of the Parties shall examine, on the basis of studies to be carried out, the issue of liability and redress, including restoration and compensation, for damage to biological diversity, except where such liability is a purely internal matter.

Article 15. Access to Genetic Resources—(1) Recognizing the sovereign rights of States over their natural resources, the authority to determine access to genetic resources rests with the national governments and is subject to national legislation.

(2) Each Contracting Party shall endeavour to create conditions to facilitate access to genetic resources for environmentally sound uses by other Contracting Parties and not to impose restrictions that run counter to the objectives of this Convention.

(3) For the purpose of this Convention, the genetic resources being provided by a Contracting Party, as referred to in this Article and Articles 16 and 19, are only those that are provided by Contracting Parties that are countries of origin of such resources or by the Parties that have acquired the genetic resources in accordance with this Convention.

(4) Access, where granted, shall be on mutually agreed terms and subject to the provisions of this Article.

(5) Access to genetic resources shall be subject to prior informed consent of the Contracting Party providing such resources, unless otherwise determined by that Party.

(6) Each Contracting Party shall endeavour to develop and carry out scientific research based on genetic resources provided by other Contracting Parties with the full participation of, and where possible in, such Contracting Parties.

(7) Each Contracting Party shall take legislative, administrative or policy measures, as appropriate, and in accordance with Articles 16 and 19 and, where necessary, through the financial mechanism established by Articles 20 and 21 with the aim of sharing in a fair and equitable way the results of research and development and the benefits arising from the commercial and other utilization of genetic resources with the Contracting Party providing such resources. Such sharing shall be upon mutually agreed terms.

Article 16. Access to and Transfer of Technology—(1) Each Contracting Party, recognizing that technology includes biotechnology, and that both access to and transfer of technology among Contracting Parties are essential elements for the attainment of the objectives of this Convention, undertakes subject to the provisions of this Article to provide and/or facilitate access for and transfer to other Contracting Parties of technologies that are relevant to the conservation and sustainable use of biological diversity or make use of genetic resources and do not cause significant damage to the environment.

(2) Access to and transfer of technology referred to in paragraph 1 above to developing countries shall be provided and/or facilitated under fair and most favourable terms, including on concessional and preferential terms where mutually agreed, and, where necessary, in accordance with the financial mechanism established by Articles 20 and 21. In the case of technology subject to patents and other intellectual property rights, such access and transfer shall be provided on terms which recognize and are consistent with the adequate and effective protection of intellectual property rights. The application of this paragraph shall be consistent with paragraphs 3, 4 and 5 below.

(3) Each Contracting Party shall take legislative, administrative

or policy measures, as appropriate, with the aim that Contracting Parties, in particular those that are developing countries, which provide genetic resources are provided access to and transfer of technology which makes use of those resources, on mutually agreed terms, including technology protected by patents and other intellectual property rights, where necessary, through the provisions of Articles 20 and 21 and in accordance with international law and consistent with paragraphs 4 and 5 below.

(4) Each Contracting Party shall take legislative, administrative or policy measures, as appropriate, with the aim that the private sector facilitates access to, joint development and transfer of technology referred to in paragraph 1 above for the benefit of both governmental institutions and the private sector of developing countries and in this regard shall abide by the obligations included in paragraphs 1, 2 and 3 above.

(5) The Contracting Parties, recognizing that patents and other intellectual property rights may have an influence on the implementation of this Convention, shall cooperate in this regard subject to national legislation and international law in order to ensure that such rights are supportive of and do not run counter to its objectives.

Article 17. Exchange of Information—(1) The Contracting Parties shall facilitate the exchange of information, from all publicly available sources, relevant to the conservation and sustainable use of biological diversity, taking into account the special needs of developing countries.

(2) Such exchange of information shall include exchange of results of technical, scientific and socio-economic research, as well as information on training and surveying programmes, specialized knowledge, indigenous and traditional knowledge as such and in combination with the technologies referred to in Article 16, paragraph 1. It shall also, where feasible, include repatriation of information.

Article 18. Technical and Scientific Cooperation—(1) The Contracting Parties shall promote international technical and scientific cooperation in the field of conservation and sustainable use of biological diversity, where necessary, through the appropriate international and national institutions.

(2) Each Contracting Party shall promote technical and scientific cooperation with other Contracting Parties, in particular developing

countries, in implementing this Convention, *inter alia*, through the development and implementation of national policies. In promoting such cooperation, special attention should be given to the development and strengthening of national capabilities, by means of human resources development and institution building.

(3) The Conference of the Parties, at its first meeting, shall determine how to establish a clearing-house mechanism to promote and facilitate technical and scientific cooperation.

(4) The Contracting Parties shall, in accordance with national legislation and policies, encourage and develop methods of cooperation for the development and use of technologies, including indigenous and traditional technologies, in pursuance of the objectives of this Convention. For this purpose, the Contracting Parties shall also promote cooperation in the training of personnel and exchange of experts.

(5) The Contracting Parties shall, subject to mutual agreement, promote the establishment of joint research programmes and joint ventures for the development of technologies relevant to the objectives of this Convention.

Article 19. Handling of Biotechnology and Distribution of its Benefits—(1) Each Contracting Party shall take legislative, administrative or policy measures, as appropriate, to provide for the effective participation in biotechnological research activities by those Contracting Parties, especially developing countries, which provide the genetic resources for such research, and where feasible in such Contracting Parties.

(2) Each Contracting Party shall take all practicable measures to promote and advance priority access on a fair and equitable basis by Contracting Parties, especially developing countries, to the results and benefits arising from biotechnologies based upon genetic resources provided by those Contracting Parties. Such access shall be on mutually agreed terms.

(3) The Parties shall consider the need for and modalities of a protocol setting out appropriate procedures, including, in particular, advance informed agreement, in the field of the safe transfer, handling and use of any living modified organism resulting from biotechnology that may have adverse effect on the conservation and sustainable use of biological diversity.

(4) Each Contracting Party shall, directly or by requiring any natural or legal person under its jurisdiction providing the

organisms referred to in paragraph 3 above, provide any available information about the use and safety regulations required by that Contracting Party in handling such organisms, as well as any available information on the potential adverse impact of the specific organisms concerned to the Contracting Party into which those organisms are to be introduced.

Article 20. Financial Resources—(1) Each Contracting Party undertakes to provide, in accordance with its capabilities, financial support and incentives in respect of those national activities which are intended to achieve the objectives of this Convention, in accordance with its national plans, priorities and programmes.

(2) The developed country Parties shall provide new and additional financial resources to enable developing country Parties to meet the agreed full incremental costs to them of implementing measures which fulfil the obligations of this Convention and to benefit from its provisions and which costs are agreed between a developing country Party and the institutional structure referred to in Article 21, in accordance with policy, strategy, programme priorities and eligibility criteria and an indicative list of incremental costs established by the Conference of the Parties. Other Parties, including countries undergoing the process of transition to a market economy, may voluntarily assume the obligations of the developed country Parties. For the purpose of this Article, the Conference of the Parties, shall at its first meeting establish a list of developed country Parties and other Parties which voluntarily assume the obligations of the developed country Parties. The Conference of the Parties shall periodically review and if necessary amend the list. Contributions from other countries and sources on a voluntary basis would also be encouraged. The implementation of these commitments shall take into account the need for adequacy, predictability and timely flow of funds and the importance of burden-sharing among the contributing Parties included in the list.

(3) The developed country Parties may also provide, and developing country Parties avail themselves of, financial resources related to the implementation of this Convention through bilateral, regional and other multilateral channels.

(4) The extent to which developing country Parties will effectively implement their commitments under this Convention

will depend on the effective implementation by developed country Parties of their commitments under this Convention related to financial resources and transfer of technology and will take fully into account the fact that economic and social development and eradication of poverty are the first and overriding priorities of the developing country Parties.

(5) The Parties shall take full account of the specific needs and special situation of least developed countries in their actions with regard to funding and transfer of technology.

(6) The Contracting Parties shall also take into consideration the special conditions resulting from the dependence on, distribution and location of, biological diversity within developing country Parties, in particular small island States.

(7) Consideration shall also be given to the special situation of developing countries, including those that are most environmentally vulnerable, such as those with arid and semi-arid zones, coastal and mountainous areas.

Article 21. Financial Mechanism—(1) There shall be a mechanism for the provision of financial resources to developing country Parties for purposes of this Convention on a grant or concessional basis the essential elements of which are described in this Article. The mechanism shall function under the authority and guidance of, and be accountable to, the Conference of the Parties for purposes of this Convention. The operations of the mechanism shall be carried out by such institutional structure as may be decided upon by the Conference of the Parties at its first meeting. For purposes of this Convention, the Conference of the Parties shall determine the policy, strategy, programme priorities and eligibility criteria relating to the access to and utilization of such resources. The contributions shall be such as to take into account the need for predictability, adequacy and timely flow of funds referred to in Article 20 in accordance with the amount of resources needed to be decided periodically by the Conference of the Parties and the importance of burden-sharing among the contributing Parties included in the list referred to in Article 20, paragraph 2. Voluntary contributions may also be made by the developed country Parties and by other countries and sources. The mechanism shall operate within a democratic and transparent system of governance.

(2) Pursuant to the objectives of this Convention, the Conference

of the Parties shall at its first meeting determine the policy, strategy and programme priorities, as well as detailed criteria and guidelines for eligibility for access to and utilization of the financial resources including monitoring and evaluation on a regular basis of such utilization. The Conference of the Parties shall decide on the arrangements to give effect to paragraph 1 above after consultation with the institutional structure entrusted with the operation of the financial mechanism.

(3) The Conference of the Parties shall review the effectiveness of the mechanism established under this Article, including the criteria and guidelines referred to in paragraph 2 above, not less than two years after the entry into force of this Convention and thereafter on a regular basis. Based on such review, it shall take appropriate action to improve the effectiveness of the mechanism if necessary.

(4) The Contracting Parties shall consider strengthening existing financial institutions to provide financial resources for the conservation and sustainable use of biological diversity.

Article 22. Relationship with Other International Conventions—(1) The provisions of this Convention shall not affect the rights and obligations of any Contracting Party deriving from any existing international agreement, except where the exercise of those rights and obligations would cause a serious damage or threat to biological diversity.

(2) Contracting Parties shall implement this Convention with respect to the marine environment consistently with the rights and obligations of States under the law of the sea.

Article 23. Conference of the Parties—(1) A Conference of the Parties is hereby established. The first meeting of the Conference of the Parties shall be convened by the Executive Director of the United Nations Environment Programme not later than one year after the entry into force of this Convention. Thereafter, ordinary meetings of the Conference of the Parties shall be held at regular intervals to be determined by the Conference at its first meeting.

(2) Extraordinary meetings of the Conference of the Parties shall be held at such other times as may be deemed necessary by the Conference, or at the written request of any Party, provided that, within six months of the request being communicated to them by the Secretariat, it is supported by at least one third of the Parties.

(3) The Conference of the Parties shall by consensus agree upon and adopt rules of procedure for itself and for any subsidiary body it may establish, as well as financial rules governing the funding of the Secretariat. At each ordinary meeting, it shall adopt a budget for the financial period until the next ordinary meeting.

(4) The Conference of the Parties shall keep under review the implementation of this Convention, and, for this purpose, shall:

- (*a*) Establish the form and the intervals for transmitting the information to be submitted in accordance with Article 26 and consider such information as well as reports submitted by any subsidiary body;
- (*b*) Review scientific, technical and technological advice on biological diversity provided in accordance with Article 25;
- (*c*) Consider and adopt, as required, protocols in accordance with Article 28;
- (*d*) Consider and adopt, as required, in accordance with Articles 29 and 30, amendments to this Convention and its annexes;
- (*e*) Consider amendments to any protocol, as well as to any annexes thereto, and, if so decided, recommend their adoption to the parties to the protocol concerned;
- (*f*) Consider and adopt, as required, in accordance with Article 30, additional annexes to this Convention;
- (*g*) Establish such subsidiary bodies, particularly to provide scientific and technical advice, as are deemed necessary for the implementation of this Convention;
- (*h*) Contact, through the Secretariat, the executive bodies of conventions dealing with matters covered by this Convention with a view to establishing appropriate forms of cooperation with them; and
- (*i*) Consider and undertake any additional action that may be required for the achievement of the purposes of this Convention in the light of experience gained in its operation.

(5) The United Nations, its specialized agencies and the International Atomic Energy Agency, as well as any State not Party to this Convention, may be represented as observers at meetings of the Conference of the Parties. Any other body or agency, whether governmental or non-governmental, qualified in fields relating to conservation and sustainable use of biological

diversity, which has informed the Secretariat of its wish to be represented as an observer at a meeting of the Conference of the Parties, may be admitted unless at least one third of the Parties present object. The admission and participation of observers shall be subject to the rules of procedure adopted by the Conference of the Parties.

Article 24. Secretariat—(1) A secretariat is hereby established. Its functions shall be:

(*a*) To arrange for and service meetings of the Conference of the Parties provided for in Article 23;

(*b*) To perform the functions assigned to it by any protocol;

(*c*) To prepare reports on the execution of its functions under this Convention and present them to the Conference of the Parties;

(*d*) To coordinate with other relevant international bodies and, in particular to enter into such administrative and contractual arrangements as may be required for the effective discharge of its functions; and

(*e*) To perform such other functions as may be determined by the Conference of the Parties.

(2) At its first ordinary meeting, the Conference of the Parties shall designate the secretariat from amongst those existing competent international organizations which have signified their willingness to carry out the secretariat functions under this Convention.

Article 25. Subsidiary Body on Scientific, Technical and Technological Advice—(1) A subsidiary body for the provision of scientific, technical and technological advice is hereby established to provide the Conference of the Parties and, as appropriate, its other subsidiary bodies with timely advice relating to the implementation of this Convention. This body shall be open to participation by all Parties and shall be multidisciplinary. It shall comprise government representatives competent in the relevant field of expertise. It shall report regularly to the Conference of the Parties on all aspects of its work.

(2) Under the authority of and in accordance with guidelines laid down by the Conference of the Parties, and upon its request, this body shall:

(*a*) Provide scientific and technical assessments of the status of biological diversity;

(*b*) Prepare scientific and technical assessments of the effects of types of measures taken in accordance with the provisions of this Convention;

(*c*) Identify innovative, efficient and state-of-the-art technologies and know-how relating to the conservation and sustainable use of biological diversity and advise on the ways and means of promoting development and/or transferring such technologies;

(*d*) Provide advice on scientific programmes and international cooperation in research and development related to conservation and sustainable use of biological diversity; and

(*e*) Respond to scientific, technical, technological and methodological questions that the Conference of the Parties and its subsidiary bodies may put to the body.

(3) The functions, terms of reference, organization and operation of this body may be further elaborated by the Conference of the Parties.

Article 26. Reports—Each Contracting Party shall, at intervals to be determined by the Conference of the Parties, present to the Conference of the Parties, reports on measures which it has taken for the implementation of the provisions of this Convention and their effectiveness in meeting the objectives of this Convention.

Article 27. Settlement of Disputes—(1) In the event of a dispute between Contracting Parties concerning the interpretation or application of this Convention, the parties concerned shall seek solution by negotiation.

(2) If the parties concerned cannot reach agreement by negotiation, they may jointly seek the good offices of, or request mediation by, a third party.

(3) When ratifying, accepting, approving or acceding to this Convention, or at any time thereafter, a State or regional economic integration organization may declare in writing to the Depositary that for a dispute not resolved in accordance with paragraph 1 or paragraph 2 above, it accepts one or both of the following means of dispute settlement as compulsory:

(*a*) Arbitration in accordance with the procedure laid down in Part 1 of Annex II;

(*b*) Submission of the dispute to the International Court of Justice.

(4) If the parties to the dispute have not, in accordance with paragraph 3 above, accepted the same or any procedure, the dispute shall be submitted to conciliation in accordance with Part 2 of Annex II unless the parties otherwise agree.

(5) The provisions of this Article shall apply with respect to any protocol except as otherwise provided in the protocol concerned.

Article 28. Adoption of Protocols—(1) The Contracting Parties shall cooperate in the formulation and adoption of protocols to this Convention.

(2) Protocols shall be adopted at a meeting of the Conference of the Parties.

(3) The text of any proposed protocol shall be communicated to the Contracting Parties by the Secretariat at least six months before such a meeting.

Article 29. Amendment of the Convention or Protocols—(1) Amendments to this Convention may be proposed by any Contracting Party. Amendments to any protocol may be proposed by any Party to that protocol.

(2) Amendments to this Convention shall be adopted at a meeting of the Conference of the Parties. Amendments to any protocol shall be adopted at a meeting of the Parties to the Protocol in question. The text of any proposed amendment to this Convention or to any protocol, except as may otherwise be provided in such protocol, shall be communicated to the Parties to the instrument in question by the secretariat at least six months before the meeting at which it is proposed for adoption. The secretariat shall also communicate proposed amendments to the signatories to this Convention for information.

(3) The Parties shall make every effort to reach agreement on any proposed amendment to this Convention or to any protocol by consensus. If all efforts at consensus have been exhausted, and no agreement reached, the amendment shall as a last resort be adopted by a two-third majority vote of the Parties to the instrument in question present and voting at the meeting, and shall be submitted by the Depositary to all Parties for ratification, acceptance or approval.

(4) Ratification, acceptance or approval of amendments shall be notified to the Depositary in writing. Amendments adopted in accordance with paragraph 3 above shall enter into force among

Parties having accepted them on the ninetieth day after the deposit of instruments of ratification, acceptance or approval by at least two thirds of the Contracting Parties to this Convention or of the Parties to the protocol concerned, except as may otherwise be provided in such protocol. Thereafter the amendments shall enter into force for any other Party on the ninetieth day after that Party deposits its instrument of ratification, acceptance or approval of the amendments.

(5) For the purposes of this Article, "Parties present and voting" means Parties present and casting an affirmative or negative vote.

Article 30. Adoption and Amendment of Annexes—(1) The annexes to this Convention or to any protocol shall form an integral part of the Convention or of such protocol, as the case may be, and, unless expressly provided otherwise, a reference to this Convention or its protocols constitutes at the same time a reference to any annexes thereto. Such annexes shall be restricted to procedural, scientific, technical and administrative matters.

(2) Except as may be otherwise provided in any protocol with respect to its annexes, the following procedure shall apply to the proposal, adoption and entry into force of additional annexes to this Convention or of annexes to any protocol:

(*a*) Annexes to this Convention or to any protocol shall be proposed and adopted according to the procedure laid down in Article 29;

(*b*) Any Party that is unable to approve an additional annex to this Convention or an annex to any protocol to which it is Party shall so notify the Depositary, in writing, within one year from the date of the communication of the adoption by the Depositary. The Depositary shall without delay notify all Parties of any such notification received. A Party may at any time withdraw a previous declaration of objection and the annexes shall thereupon enter into force for that Party subject to subparagraph (*c*) below;

(*c*) On the expiry of one year from the date of the communication of the adoption by the Depositary, the annex shall enter into force for all Parties to this Convention or to any protocol concerned which have not submitted a notification in accordance with the provisions of subparagraph (*b*) above.

(3) The proposal, adoption and entry into force of amendments to annexes to this Convention or to any protocol shall be subject to the same procedure as for the proposal, adoption and entry into force of annexes to the Convention or annexes to any protocol.

(4) If an additional annex or an amendment to an annex is related to an amendment to this Convention or to any protocol, the additional annex or amendment shall not enter into force until such time as the amendment to the Convention or to the protocol concerned enters into force.

Article 31. Right to Vote—(1) Except as provided for in paragraph 2 below, each Contracting Party to this Convention or to any protocol shall have one vote.

(2) Regional economic integration organizations, in matters within their competence, shall exercise their right to vote with a number of votes equal to the number of their member States which are Contracting Parties to this Convention or the relevant protocol. Such organizations shall not exercise their right to vote if their member States exercise theirs, and vice versa.

Article 32. Relationship between this Convention and Its Protocols—(1) A State or a regional economic integration organization may not become a Party to a protocol unless it is, or becomes at the same time, a Contracting Party to this Convention.

(2) Decisions under any protocol shall be taken only by the Parties to the protocol concerned. Any Contracting Party that has not ratified, accepted or approved a protocol may participate as an observer in any meeting of the parties to that protocol.

Article 33. Signature—This Convention shall be open for signature at Rio de Janeiro by all States and any regional economic integration organization from 5 June 1992 until 14 June 1992, and at the United Nations Headquarters in New York from 15 June 1992 to 4 June 1993.

Article 34. Ratification, Acceptance or Approval—(1) This Convention and any protocol shall be subject to ratification, acceptance or approval by States and by regional economic integration organizations. Instruments of ratification, acceptance or approval shall be deposited with the Depositary.

(2) Any organization referred to in paragraph 1 above which becomes a Contracting Party to this Convention or any protocol without any of its member States being a Contracting Party shall be bound by all the obligations under the Convention or the

protocol, as the case may be. In the case of such organizations, one or more of whose member States is a Contracting Party to this Convention or relevant protocol, the organization and its member States shall decide on their respective responsibilities for the performance of their obligations under the Convention or protocol, as the case may be. In such cases, the organization and the member States shall not be entitled to exercise rights under the Convention or relevant protocol concurrently.

(3) In their instruments of ratification, acceptance or approval, the organizations referred to in paragraph 1 above shall declare the extent of their competence with respect to the matters governed by the Convention or the relevant protocol. These organizations shall also inform the Depositary of any relevant modification in the extent of their competence.

Article 35. Accession—(1) This Convention and any protocol shall be open for accession by States and by regional economic integration organizations from the date on which the Convention or the protocol concerned is closed for signature. The instruments of accession shall be deposited with the Depositary.

(2) In their instruments of accession, the organizations referred to in paragraph 1 above shall declare the extent of their competence with respect to the matters governed by the Convention or the relevant protocol. These organizations shall also inform the Depositary of any relevant modification in the extent of their competence.

(3) The provisions of Article 34, paragraph 2, shall apply to regional economic integration organizations which accede to this Convention or any protocol.

Article 36. Entry Into Force—(1) This Convention shall enter into force on the ninetieth day after the date of deposit of the thirtieth instrument of ratification, acceptance, approval or accession.

(2) Any protocol shall enter into force on the ninetieth day after the date of deposit of the number of instruments of ratification, acceptance, approval or accession, specified in that protocol, has been deposited.

(3) For each Contracting Party which ratifies, accepts or approves this Convention or accedes thereto after the deposit of the thirtieth instrument of ratification, acceptance, approval or accession, it shall enter into force on the ninetieth day after the

date of deposit by such Contracting Party of its instrument of ratification, acceptance, approval or accession.

(4) Any protocol, except as otherwise provided in such protocol, shall enter into force for a Contracting Party that ratifies, accepts or approves that protocol or accedes thereto after its entry into force pursuant to paragraph 2 above, on the ninetieth day after the date on which that Contracting Party deposits its instrument of ratification, acceptance, approval or accession, or on the date on which this Convention enters into force for that Contracting Party, whichever shall be the later.

(5) For the purposes of paragraphs 1 and 2 above, any instrument deposited by a regional economic integration organization shall not be counted as additional to those deposited by member States of such organization.

Article 37. Reservations—No reservations may be made to this Convention.

Article 38. Withdrawals—(1) At any time after two years from the date on which this Convention has entered into force for a Contracting Party, that Contracting Party may withdraw from the Convention by giving written notification to the Depositary.

(2) Any such withdrawal shall take place upon expiry of one year after the date of its receipt by the Depositary, or on such later date as may be specified in the notification of the withdrawal.

(3) Any Contracting Party which withdraws from this Convention shall be considered as also having withdrawn from any protocol to which it is party.

Article 39. Financial Interim Arrangements—Provided that it has been fully restructured in accordance with the requirements of Article 21, the Global Environment Facility of the United Nations Development Programme, the United Nations Environment Programme and the International Bank for Reconstruction and Development shall be the institutional structure referred to in Article 21 on an interim basis, for the period between the entry into force of this Convention and the first meeting of the Conference of the Parties or until the Conference of the Parties decides which institutional structure will be designated in accordance with Article 21.

Article 40. Secretariat Interim Arrangements—The secretariat to be provided by the Executive Director of the United Nations Environment Programme shall be the secretariat referred to in Article 24, paragraph 2, on an interim basis for the period between the entry into force of this Convention and the first meeting of the Conference of the Parties.

Article 41. Depositary—The Secretary-General of the United Nations shall assume the functions of Depositary of this Convention and any protocols.

Article 42. Authentic Texts—The original of this Convention, of which the Arabic, Chinese, English, French, Russian and Spanish texts are equally authentic, shall be deposited with the Secretary-General of the United Nations.

In witness whereof the undersigned, being duly authorized to that effect, have signed this Convention.

Done at Rio de Janeiro on this fifth day of June, one thousand nine hundred and ninety-two.

Annex I
Identification and Monitoring

1. Ecosystems and habitats: containing high diversity, large numbers of endemic or threatened species, or wilderness; required by migratory species; of social, economic, cultural or scientific importance; or, which are representative, unique or associated with key evolutionary or other biological processes;
2. Species and communities which are: threatened; wild relatives of domesticated or cultivated species; of medicinal, agricultural or other economic value; or social, scientific or cultural importance; or importance for research into the conservation and sustainable use of biological diversity, such as indicator species; and
3. Described genomes and genes of social, scientific or economic importance.

Annex II

Part 1: Arbitration

Article 1. The claimant party shall notify the secretariat that the parties are referring a dispute to arbitration pursuant to Article 27. The notification shall state the subject-matter of arbitration and include, in particular, the articles of the Convention or the protocol, the interpretation or application of which are at issue. If the parties do not agree on the subject matter of the dispute before the President of the tribunal is designated, the arbitral tribunal shall determine the subject matter. The secretariat shall forward the information thus received to all Contracting Parties to this Convention or to the protocol concerned.

Article 2. (1) In disputes between two parties, the arbitral tribunal shall consist of three members. Each of the parties to the dispute shall appoint an arbitrator and the two arbitrators so appointed shall designate by common agreement the third arbitrator who shall be the President of the tribunal. The latter shall not be a national of one of the parties to the dispute, nor have his or her usual place of residence in the territory of one of these parties, nor be employed by any of them, nor have dealt with the case in any other capacity.

(2) In disputes between more than two parties, parties in the same interest shall appoint one arbitrator jointly by agreement.

(3) Any vacancy shall be filled in the manner prescribed for the initial appointment.

Article 3. (1) If the President of the arbitral tribunal has not been designated within two months of the appointment of the second arbitrator, the Secretary-General of the United Nations shall, at the request of a party, designate the President within a further two-month period.

(2) If one of the parties to the dispute does not appoint an arbitrator within two months of receipt of the request, the other party may inform the Secretary-General who shall make the designation within a further two-month period.

Article 4. The arbitral tribunal shall render its decisions in accordance with the provisions of this Convention, any protocols concerned, and international law.

Article 5. Unless the parties to the dispute otherwise agree, the arbitral tribunal shall determine its own rules of procedure.

Article 6. The arbitral tribunal may, at the request of one of the parties, recommend essential interim measures of protection.

Article 7. The parties to the dispute shall facilitate the work of the arbitral tribunal and, in particular, using all means at their disposal, shall:

(*a*) Provide it with all relevant documents, information and facilities; and

(*b*) Enable it, when necessary, to call witnesses or experts and receive their evidence.

Article 8. The parties and the arbitrators are under an obligation to protect the confidentiality of any information they receive in confidence during the proceedings of the arbitral tribunal.

Article 9. Unless the arbitral tribunal determines otherwise because of the particular circumstances of the case, the costs of the tribunal shall be borne by the parties to the dispute in equal shares. The tribunal shall keep a record of all its costs, and shall furnish a final statement thereof to the parties.

Article 10. Any Contracting Party that has an interest of a legal nature in the subject-matter of the dispute which may be affected by the decision in the case, may intervene in the proceedings with the consent of the tribunal.

Article 11. The tribunal may hear and determine counterclaims arising directly out of the subject-matter of the dispute.

Article 12. Decisions both on procedure and substance of the arbitral tribunal shall be taken by a majority vote of its members.

Article 13. If one of the parties to the dispute does not appear before the arbitral tribunal or fails to defend its case, the other party may request the tribunal to continue the proceedings and to make its award. Absence of a party or a failure of a party to defend its case shall not constitute a bar to the proceedings. Before rendering its final decision, the arbitral tribunal must satisfy itself that the claim is well founded in fact and law.

Article 14. The tribunal shall render its final decision within five months of the date on which it is fully constituted unless it finds it necessary to extend the time-limit for a period which should not exceed five more months.

Article 15. The final decision of the arbitral tribunal shall be confined to the subject-matter of the dispute and shall state the reasons on which it is based. It shall contain the names of the members who have participated and the date of the final decision. Any member of the tribunal may attach a separate or dissenting opinion to the final decision.

Article 16. The award shall be binding on the parties to the dispute. It shall be without appeal unless the parties to the dispute have agreed in advance to an appellate procedure.

Article 17. Any controversy which may arise between the parties to the dispute as regards the interpretation or manner of implementation of the final decision may be submitted by either party for decision to the arbitral tribunal which rendered it.

Part 2: Conciliation

Article 1. A conciliation commission shall be created upon the request of one of the parties to the dispute. The commission shall, unless the parties otherwise agree, be composed of five members, two appointed by each Party concerned and a President chosen jointly by those members.

Article 2. In disputes between more than two parties, parties in the same interest shall appoint their members of the commission jointly by agreement. Where two or more parties have separate interests or there is a disagreement as to whether they are of the same interest, they shall appoint their members separately.

Article 3. If any appointments by the parties are not made within two months of the date of the request to create a conciliation commission, the Secretary-General of the United Nations shall, if asked to do so by the party that made the request, make those appointments within a further two-month period.

Article 4. If a President of the conciliation commission has not been chosen within two months of the last of the members of the commission being appointed, the Secretary-General of the United Nations shall, if asked to do so by a party, designate a President within a further two-month period.

Article 5. The conciliation commission shall take its decisions by majority vote of its members. It shall, unless the parties to the dispute otherwise agree, determine its own procedure. It shall render a proposal for resolution of the dispute, which the parties shall consider in good faith.

Article 6. A disagreement as to whether the conciliation commission has competence shall be decided by the commission.

PARTIES TO THE CONVENTION ON BIOLOGICAL DIVERSITY

Parties Information
Sort by CBD ratification date

Convention on Biological Diversity: 188 Parties (168 Signatures)
Cartagena Protocol on Biosafety: 131 Parties (103 Signatures)

**Note*: rtf + Ratification, acs = Accession, acp = Acceptance, apv = approval, Date format: dd/mm/yyyy

Country		*Convention on Biological Diversity*		*Biosafety Protocol*		
Code	*Name*	*Signed*	*Party**	*Signed*	*Ratification**	*Party*
af	Afghanistan	12/06/1992	19/09/2002 rtf			
al	Albania		05/01/1994 acs		08/02/2005 acs	09/05/2005
dz	Algeria	13/06/1992	14/08/1995 rtf	25/05/2000	05/08/2004 rtf	03/11/2004
ad	Andorra					
ao	Angola	12/06/1992	01/04/1998 rtf			
ag	Antigua and Barbuda	05/06/1992	09/03/1993 rtf	24/05/2000	10/09/2003 rtf	09/12/2003
ar	Argentina	12/06/1992	22/11/1994 rtf	24/05/2000		
am	Armenia	13/06/1992	14/05/1993 acp		30/04/2004 acs	29/07/2004
au	Australia	05/06/1992	18/06/1993 rtf			
at	Austria	13/06/1992	18/08/1994 rtf	24/05/2000	27/08/2002 rtf	11/09/2003
az	Azerbaijan	12/06/1992	03/08/2000 apv		01/04/2005 acs	30/06/2005

(Contd.)

(*Contd.*)

bs	Bahamas	12/06/1992	02/09/1993 rtf	24/05/2000	15/01/2004 rtf	14/04/2004
bh	Bahrain	09/06/1992	30/08/1996 rtf			
bd	Bangladesh	05/06/1992	03/05/1994 rtf	24/05/2000	05/02/2004 rtf	05/05/2004
bb	Barbados	12/06/1992	10/12/1993 rtf		06/09/2002 acs	11/09/2003
by	Belarus	11/06/1992	08/09/1993 rtf		26/08/2002 acs	11/09/2003
be	Belgium	05/06/1992	22/11/1996 rtf	24/05/2000	15/04/2004 rtf	14/07/2004
bz	Belize	13/06/1992	30/12/1993 rtf		12/02/2004 acs	12/05/2004
bj	Benin	13/06/1992	30/06/1994 rtf	24/05/2000	02/03/2005 rtf	31/05/2005
bt	Bhutan	11/06/1992	25/08/1995 rtf		26/08/2002 acs	11/09/2003
bo	Bolivia	13/06/1992	03/10/1994 rtf	24/05/2000	22/04/2002 rtf	11/09/2003
ba	Bosnia and Herzegovina		26/08/2002 acs			
bw	Botswana	08/06/1992	12/10/1995 rtf	01/06/2001	11/06/2002 rtf	11/09/2003
br	Brazil	05/06/1992	28/02/1994 rtf		24/11/2003 acs	22/02/2004
bn	Brunei Darussalam					
bg	Bulgaria	12/06/1992	17/04/1996 rtf	24/05/2000	13/10/2000 rtf	11/09/2003
bf	Burkina Faso	12/06/1992	02/09/1993 rtf	24/05/2000	04/08/2003 rtf	02/11/2003
bi	Burundi	11/06/1992	15/04/1997 rtf			
kh	Cambodia		09/02/1995 acs		17/09/2003 acs	16/12/2003
cm	Cameroon	14/06/1992	19/10/1994 rtf	09/02/2001	20/02/2003 rtf	11/09/2003
ca	Canada	11/06/1992	04/12/1992 rtf	19/04/2001		
cv	Cape Verde	12/06/1992	29/03/1995 rtf		01/11/2005 acs	30/01/2006
cf	Central African Republic	13/06/1992	15/03/1995 rtf	24/05/2000		
td	Chad	12/06/1992	07/06/1994 rtf	24/05/2000		
cl	Chile	13/06/1992	09/09/1994 rtf	24/05/2000		

(*Contd.*)

(*Contd.*)

cn	China	11/06/1992	05/01/1993 rtf	08/08/2000	08/06/2005 apv	06/09/2005
co	Colombia	12/06/1992	28/11/1994 rtf	24/05/2000	20/05/2003 rtf	11/09/2003
km	Comoros	11/06/1992	29/09/1994 rtf			
cg	Congo	11/06/1992	01/08/1996 rtf	21/11/2000		
ck	Cook Islands	12/06/1992	20/04/1993 rtf	21/05/2001		
cr	Costa Rica	13/06/1992	26/08/1994 rtf	24/05/2000		
ci	Côte d'Ivoire	10/06/1992	29/11/1994 rtf			
hr	Croatia	11/06/1992	07/10/1996 rtf	08/09/2000	29/08/2002 rtf	11/09/2003
cu	Cuba	12/06/1992	08/03/1994 rtf	24/05/2000	17/09/2002 rtf	11/09/2003
cy	Cyprus	12/06/1992	10/07/1996 rtf		05/12/2003 acs	04/03/2004
cz	Czech Republic	04/06/1993	03/12/1993 apv	24/05/2000	08/10/2001 rtf	11/09/2003
kp	Democratic People's Republic of Korea	11/06/1992	26/10/1994 apv	20/04/2001	29/07/2003 rtf	27/10/2003
cd	Democratic Republic of the Congo	11/06/1992	03/12/1994 rtf		23/03/2005 acs	21/06/2005
dk	Denmark	12/06/1992	21/12/1993 rtf	24/05/2000	27/08/2002 rtf	11/09/2003
dj	Djibouti	13/06/1992	01/09/1994 rtf		08/04/2002 acs	11/09/2003
dm	Dominica		06/04/1994 rtf		13/07/2004 acs	11/10/2004
do	Dominican Republic	13/06/1992	25/11/1996 rtf			
ec	Ecuador	09/06/1992	23/02/1993 rtf	24/05/2000	30/01/2003 rtf	11/09/2003
eg	Egypt	09/06/1992	02/06/1994 rtf	20/12/2000	23/12/2003 rtf	21/03/2004
sv	El Salvador	13/06/1992	08/09/1994 rtf	24/05/2000	26/09/2003 rtf	25/12/2003
gq	Equatorial Guinea		06/12/1994 acs			
er	Eritrea		21/03/1996 acs		10/03/2005 acs	08/06/2005

(*Contd.*)

(*Contd.*)

ee	Estonia	12/06/1992	27/07/1994 rtf	06/09/2000	24/03/2004 rtf	22/06/2004
et	Ethiopia	10/06/1992	05/04/1994 rtf	24/05/2000	09/10/2003 rtf	07/01/2004
eur	European Community	13/06/1992	21/12/1993 apv	24/05/2000	apv	11/09/2003
fj	Fiji	09/10/1992	25/02/1993 rtf	02/05/2001	05/06/2001 rtf	11/09/2003
fi	Finland	05/06/1992	27/07/1994 acp	24/05/2000	09/07/2004 rtf	07/10/2004
fr	France	13/06/1992	01/07/1994 rtf	24/05/2000	07/04/2003 apv	11/09/2003
ga	Gabon	12/06/1992	14/03/1997 rtf			
gm	Gambia	12/06/1992	10/06/1994 rtf	24/05/2000	09/06/2004 rtf	07/09/2004
ge	Georgia		02/06/1994 acs			
de	Germany	12/06/1992	21/12/1993 rtf	24/05/2000	20/11/2003 rtf	18/02/2004
gh	Ghana	12/06/1992	29/08/1994 rtf		30/05/2003 acs	11/09/2003
gr	Greece	12/06/1992	04/08/1994 rtf	24/05/2000	21/05/2004 rtf	19/08/2004
gd	Grenada	03/12/1992	11/08/1994 rtf	24/05/2000	05/02/2004 rtf	05/05/2004
gt	Guatemala	13/06/1992	10/07/1995 rtf		28/10/2004 acs	26/01/2005
gn	Guinea	12/06/1992	07/05/1993 rtf	24/05/2000		
gw	Guinea-Bissau	12/06/1992	27/10/1995 rtf			
gy	Guyana	13/06/1992	29/08/1994 rtf			
ht	Haiti	13/06/1992	25/09/1996 rtf	24/05/2000		
va	Holy See					
hn	Honduras	13/06/1992	31/07/1995 rtf	24/05/2000		
hu	Hungary	13/06/1992	24/02/1994 rtf	24/05/2000	13/01/2004 rtf	12/04/2004
is	Iceland	10/06/1992	12/09/1994 rtf	01/06/2001		
in	India	05/06/1992	18/02/1994 rtf	23/01/2001	17/01/2003 rtf	11/09/2003
id	Indonesia	05/06/1992	23/08/1994 rtf	24/05/2000	03/12/2004 rtf	03/03/2005

(*Contd.*)

(*Contd.*)

ir	Iran (Islamic Republic of)	14/06/1992	06/08/1996 rtf	23/04/2001	20/11/2003 rtf	18/02/2004
iq	Iraq					
ie	Ireland	13/06/1992	22/03/1996 rtf	24/05/2000	14/11/2003 rtf	12/02/2004
il	Israel	11/06/1992	07/08/1995 rtf			
it	Italy	05/06/1992	15/04/1994 rtf	24/05/2000	24/03/2004 rtf	22/06/2004
jm	Jamaica	11/06/1992	06/01/1995 rtf	04/06/2001		
jp	Japan	13/06/1992	28/05/1993 acp		21/11/2003 acs	19/02/2004
jo	Jordan	11/06/1992	12/11/1993 rtf	11/10/2000	11/11/2003 rtf	09/02/2004
kz	Kazakhstan	09/06/1992	06/09/1994 rtf			
ke	Kenya	11/06/1992	26/07/1994 rtf	15/05/2000	24/01/2002 rtf	11/09/2003
ki	Kiribati		16/08/1994 acs	07/09/2000	20/04/2004 rtf	19/07/2004
kw	Kuwait	09/06/1992	02/08/2002 rtf			
kg	Kyrgyzstan		06/08/1996 acs		05/10/2005 acs	03/01/2006
la	Lao People's Democratic Republic		20/09/1996 acs		03/08/2004 acs	01/11/2004
lv	Latvia	11/06/1992	14/12/1995 rtf		13/02/2004 acs	13/05/2004
lb	Lebanon	12/06/1992	15/12/1994 rtf			
ls	Lesotho	11/06/1992	10/01/1995 rtf		20/09/2001 acs	11/09/2003
lr	Liberia	12/06/1992	08/11/2000 rtf		15/02/2002 acs	11/09/2003
ly	Libyan Arab Jamahiriya	29/06/1992	12/07/2001 rtf		14/06/2005 acs	12/09/2005
li	Liechtenstein	05/06/1992	19/11/1997 rtf			
lt	Lithuania	11/06/1992	01/02/1996 rtf	24/05/2000	07/11/2003 rtf	05/02/2004
lu	Luxembourg	09/06/1992	09/05/1994 rtf	11/07/2000	28/08/2002 rtf	11/09/2003
mg	Madagascar	08/06/1992	04/03/1996 rtf	14/09/2000	24/11/2003 rtf	22/02/2004

(*Contd.*)

(Contd.)

mw	Malawi	10/06/1992	02/02/1994 rtf	24/05/2000		
my	Malaysia	12/06/1992	24/06/1994 rtf	24/05/2000	03/09/2003 rtf	02/12/2003
mv	Maldives	12/06/1992	09/11/1992 rtf		02/09/2002 acs	11/09/2003
ml	Mali	30/09/1992	29/03/1995 rtf	04/04/2001	28/08/2002 rtf	11/09/2003
mt	Malta	12/06/1992	29/12/2000 rtf			
mh	Marshall Islands	12/06/1992	08/10/1992 rtf		27/01/2003 acs	11/09/2003
mr	Mauritania	12/06/1992	16/08/1996 rtf		22/07/2005 acs	20/10/2005
mu	Mauritius	10/06/1992	04/09/1992 rtf		11/04/2002 acs	11/09/2003
mx	Mexico	13/06/1992	11/03/1993 rtf	24/05/2000	27/08/2002 rtf	11/09/2003
fm	Micronesia (Federated States of)	12/06/1992	20/06/1994 rtf			
mc	Monaco	11/06/1992	20/11/1992 rtf	24/05/2000		
mn	Mongolia	12/06/1992	30/09/1993 rtf		22/07/2003 acs	20/10/2003
ma	Morocco	13/06/1992	21/08/1995 rtf	25/05/2000		
mz	Mozambique	12/06/1992	25/08/1995 rtf	24/05/2000	21/10/2002 rtf	11/09/2003
mm	Myanmar	11/06/1992	25/11/1994 rtf	11/05/2001		
na	Namibia	12/06/1992	16/05/1997 rtf	24/05/2000	10/02/2005 rtf	11/05/2005
nr	Nauru	05/06/1992	11/11/1993 rtf		12/11/2001 acs	11/09/2003
np	Nepal	12/06/1992	23/11/1993 rtf	02/03/2001		
nl	Netherlands	05/06/1992	12/07/1994 acp	24/05/2000	08/01/2002 acp	11/09/2003
nz	New Zealand	12/06/1992	16/09/1993 rtf	24/05/2000	24/02/2005 rtf	25/05/2005
ni	Nicaragua	13/06/1992	20/11/1995 rtf	26/05/2000	28/08/2002 rtf	11/09/2003
ne	Niger	11/06/1992	25/07/1995 rtf	24/05/2000	30/09/2004 rtf	29/12/2004

(Contd.)

(Contd.)

ng	Nigeria	13/06/1992	29/08/1994 rtf	24/05/2000	15/07/2003 rtf	13/10/2003
nu	Niue		28/02/1996 acs		08/07/2002 acs	11/09/2003
no	Norway	09/06/1992	09/07/1993 rtf	24/05/2000	10/05/2001 rtf	11/09/2003
om	Oman	10/06/1992	08/02/1995 rtf		11/04/2003 acs	11/09/2003
pk	Pakistan	05/06/1992	26/07/1994 rtf	04/06/2001		
pw	Palau		06/01/1999 acs	29/05/2001	13/06/2003 rtf	11/09/2003
pa	Panama	13/06/1992	17/01/1995 rtf	11/05/2001	01/05/2002 rtf	11/09/2003
pg	Papua New Guinea	13/06/1992	16/03/1993 rtf		14/10/2005 acs	12/01/2006
py	Paraguay	12/06/1992	24/02/1994 rtf	03/05/2001	10/03/2004 rtf	08/06/2004
pe	Peru	12/06/1992	07/06/1993 rtf	24/05/2000	14/04/2004 rtf	13/07/2004
ph	Philippines	12/06/1992	08/10/1993 rtf	24/05/2000		
pl	Poland	05/06/1992	18/01/1996 rtf	24/05/2000	10/12/2003 rtf	09/03/2004
pt	Portugal	13/06/1992	21/12/1993 rtf	24/05/2000	30/09/2004 acp	29/12/2004
qa	Qatar	11/06/1992	21/08/1996 rtf			
kr	Republic of Korea	13/06/1992	03/10/1994 rtf	06/09/2000		
md	Republic of Moldova	05/06/1992	20/10/1995 rtf	14/02/2001	04/03/2003 rtf	11/09/2003
ro	Romania	05/06/1992	17/08/1994 rtf	11/10/2000	30/06/2003 rtf	28/09/2003
ru	Russian Federation	13/06/1992	05/04/1995 rtf			
rw	Rwanda	10/06/1992	29/05/1996 rtf	24/05/2000	22/07/2004 rtf	20/10/2004
kn	Saint Kitts and Nevis	12/06/1992	07/01/1993 rtf		23/05/2001 acs	11/09/2003
lc	Saint Lucia		28/07/1993 acs		16/06/2005 acs	14/09/2005
vc	Saint Vincent and the Grenadines		03/06/1996 acs		27/08/2003 acs	25/11/2003
ws	Samoa	12/06/1992	09/02/1994 rtf	24/05/2000	30/05/2002 rtf	11/09/2003

(Contd.)

(*Contd.*)

sm	San Marino	10/06/1992	28/10/1994 rtf			
st	Sao Tome and Principe	12/06/1992	29/09/1999 rtf			
sa	Saudi Arabia		03/10/2001 acs			
sn	Senegal	13/06/1992	17/10/1994 rtf	31/10/2000	08/10/2003 rtf	06/01/2004
cs	Serbia and Montenegro	08/06/1992	01/03/2002 rtf			
sc	Seychelles	10/06/1992	22/09/1992 rtf	23/01/2001	13/05/2004 rtf	11/08/2004
sl	Sierra Leone		12/12/1994 acs			
sg	Singapore	12/06/1992	21/12/1995 rtf			
sk	Slovakia	19/05/1993	25/08/1994 apv	24/05/2000	24/11/2003 rtf	22/02/2004
si	Slovenia	13/06/1992	09/07/1996 rtf	24/05/2000	20/11/2002 rtf	11/09/2003
sb	Solomon Islands	13/06/1992	03/10/1995 rtf		28/07/2004 acs	26/10/2004
so	Somalia					
za	South Africa	04/06/1993	02/11/1995 rtf		14/08/2003 acs	12/11/2003
es	Spain	13/06/1992	21/12/1993 rtf	24/05/2000	16/01/2002 rtf	11/09/2003
lk	Sri Lanka	10/06/1992	23/03/1994 rtf	24/05/2000	28/04/2004 rtf	26/07/2004
sd	Sudan	09/06/1992	30/10/1995 rtf		13/06/2005 acs	11/09/2005
sr	Suriname	13/06/1992	12/01/1996 rtf			
sz	Swaziland	12/06/1992	09/11/1994 rtf		13/01/2006 acs	13/04/2006
se	Sweden	08/06/1992	16/12/1993 rtf	24/05/2000	08/08/2002 rtf	11/09/2003
ch	Switzerland	12/06/1992	21/11/1994 rtf	24/05/2000	26/03/2002 rtf	11/09/2003
sy	Syrian Arab Republic	03/05/1993	04/01/1996 rtf		01/04/2004 acs	30/06/2004
tj	Tajikistan		29/10/1997 acs		12/02/2004 acs	12/05/2004
th	Thailand	12/06/1992	29/01/2004 rtf		10/11/2005 acs	08/02/2006
mk	The Former Yugoslav Republic of Macedonia		02/12/1997 acs	26/07/2000	14/06/2005 rtf	12/09/2005

(*Contd.*)

(Contd.)

tl	Timor-Leste					
tg	Togo	12/06/1992	04/10/1995 acp	24/05/2000	02/07/2004 rtf	30/09/2004
to	Tonga		19/05/1998 acs		18/09/2003 acs	17/12/2003
tt	Trinidad and Tobago	11/06/1992	01/08/1996 rtf		05/10/2000 acs	11/09/2003
tn	Tunisia	13/06/1992	15/07/1993 rtf	19/04/2001	22/01/2003 rtf	11/09/2003
tr	Turkey	11/06/1992	14/02/1997 rtf	24/05/2000	24/10/2003 rtf	24/01/2004
tm	Turkmenistan		18/09/1996 acs			
tv	Tuvalu	08/06/1992	20/12/2002 rtf			
ug	Uganda	12/06/1992	08/09/1993 rtf	24/05/2000	30/11/2001 rtf	11/09/2003
ua	Ukraine	11/06/1992	07/02/1995 rtf		06/12/2002 acs	11/09/2003
ae	United Arab Emirates	11/06/1992	10/02/2000 rtf			
gb	United Kingdom of Great Britain and Northern Ireland	12/06/1992	03/06/1994 rtf	24/05/2000	19/11/2003 rtf	17/02/2004
tz	United Republic of Tanzania	12/06/1992	08/03/1996 rtf		24/04/2003 acs	11/09/2003
us	United States of America	04/06/1993				
uy	Uruguay	09/06/1992	05/11/1993 rtf	01/06/2001		
uz	Uzbekistan		19/07/1995 acs			
vu	Vanuatu	09/06/1992	25/03/1993 rtf			
ve	Venezuela	12/06/1992	13/09/1994 rtf	24/05/2000	13/05/2002 rtf	11/09/2003
vn	Viet Nam	28/05/1993	16/11/1994 rtf		21/01/2004 acs	20/04/2004
ye	Yemen	12/06/1992	21/02/1996 rtf		01/12/2005 acs	01/03/2006
zm	Zambia	11/06/1992	28/05/1993 rtf		27/04/2004 acs	25/07/2004
zw	Zimbabwe	12/06/1992	11/11/1994 rtf	04/06/2001	25/02/2005 rtf	26/05/2005

CARTAGENA PROTOCOL ON BIOSAFETY

The Parties to this Protocol.

Being Parties to the Convention on Biological Diversity, hereinafter referred to as "the Convention".

Recalling Article 19, paragraphs 3 and 4, and Articles 8(*g*) and 17 of the Convention.

Recalling also decision II/5 of 17 November 1995 of the Conference of the Parties to the Convention to develop a Protocol on biosafety, specifically focusing on transboundary movement of any living modified organism resulting from modern biotechnology that may have adverse effect on the conservation and sustainable use of biological diversity, setting out for consideration, in particular, appropriate procedures for advance informed agreement.

Reaffirming the precautionary approach contained in Principle 15 of the Rio Declaration on Environment and Development.

Aware of the rapid expansion of modern biotechnology and the growing public concern over its potential adverse effects on biological diversity, taking also into account risks to human health.

Recognizing that modern biotechnology has great potential for human well-being if developed and used with adequate safety measures for the environment and human health.

Recognizing also the crucial importance to humankind of centres of origin and centres of genetic diversity.

Taking into account the limited capabilities of many countries, particularly developing countries, to cope with the nature and scale of known and potential risks associated with living modified organisms.

Recognizing that trade and environment agreements should be mutually supportive with a view to achieving sustainable development.

Emphasizing that this Protocol shall not be interpreted as implying a change in the rights and obligations of a Party under any existing international agreements.

Understanding that the above recital is not intended to subordinate this Protocol to other international agreements.

Have agreed as follows:

Article 1. Objective—In accordance with the precautionary approach contained in Principle 15 of the Rio Declaration on Environment and Development, the objective of this Protocol is to contribute to ensuring an adequate level of protection in the field of the safe transfer, handling and use of living modified organisms resulting from modern biotechnology that may have adverse effects on the conservation and sustainable use of biological diversity, taking also into account risks to human health, and specifically focusing on transboundary movements.

Article 2. General Provisions—(1) Each Party shall take necessary and appropriate legal, administrative and other measures to implement its obligations under this Protocol.

(2) The Parties shall ensure that the development, handling, transport, use, transfer and release of any living modified organisms are undertaken in a manner that prevents or reduces the risks to biological diversity, taking also into account risks to human health.

(3) Nothing in this Protocol shall affect in any way the sovereignty of States over their territorial sea established in accordance with international law, and the sovereign rights and the jurisdiction which States have in their exclusive economic zones and their continental shelves in accordance with international law, and the exercise by ships and aircraft of all States of navigational rights and freedoms as provided for in international law and as reflected in relevant international instruments.

(4) Nothing in this Protocol shall be interpreted as restricting the right of a Party to take action that is more protective of the conservation and sustainable use of biological diversity than that called for in this Protocol, provided that such action is consistent with the objective and the provisions of this Protocol and is in accordance with that Party's other obligations under international law.

(5) The Parties are encouraged to take into account, as appropriate, available expertise, instruments and work undertaken in international forums with competence in the area of risks to human health.

Article 3. Use of Terms—For the purposes of this Protocol:

(*a*) "Conference of the Parties" means the Conference of the Parties to the Convention;

(*b*) "Contained use" means any operation, undertaken within a facility, installation or other physical structure, which involves living modified organisms that are controlled by specific measures that effectively limit their contact with, and their impact on, the external environment;

(*c*) "Export" means intentional transboundary movement from one Party to another Party;

(*d*) "Exporter" means any legal or natural person, under the jurisdiction of the Party of export, who arranges for a living modified organism to be exported;

(*e*) "Import" means intentional transboundary movement into one Party from another Party;

(*f*) "Importer" means any legal or natural person, under the jurisdiction of the Party of import, who arranges for a living modified organism to be imported;

(*g*) "Living modified organism" means any living organism that possesses a novel combination of genetic material obtained through the use of modern biotechnology;

(*h*) "Living organism" means any biological entity capable of transferring or replicating genetic material, including sterile organisms, viruses and viroids;

(*i*) "Modern biotechnology" means the application of:

(*i*) *In vitro* nucleic acid techniques, including recombinant deoxyribonucleic acid (DNA) and direct injection of nucleic acid into cells or organelles, or

(*ii*) Fusion of cells beyond the taxonomic family, that overcome natural physiological reproductive or recombination barriers and that are not techniques used in traditional breeding and selection;

(*j*) "Regional economic integration organization" means an organization constituted by sovereign States of a given region, to which its member States have transferred competence in respect of matters governed by this Protocol and which has been duly authorized, in accordance with its internal procedures, to sign, ratify, accept, approve or accede to it;

(*k*) "Transboundary movement" means the movement of a living modified organism from one Party to another Party, save that for the purposes of Articles 17 and 24 transboundary movement extends to movement between Parties and non-Parties.

Article 4. Scope—This Protocol shall apply to the transboundary movement, transit, handling and use of all living modified organisms that may have adverse effects on the conservation and sustainable use of biological diversity, taking also into account risks to human health.

Article 5. Pharmaceuticals—Notwithstanding Article 4 and without prejudice to any right of a Party to subject all living modified organisms to risk assessment prior to the making of decisions on import, this Protocol shall not apply to the transboundary movement of living modified organisms which are pharmaceuticals for humans that are addressed by other relevant international agreements or organisations.

Article 6. Transit and Contained use—(1) Notwithstanding Article 4 and without prejudice to any right of a Party of transit to regulate the transport of living modified organisms through its territory and make available to the Biosafety Clearing-House, any decision of that Party, subject to Article 2, paragraph 3, regarding the transit through its territory of a specific living modified organism, the provisions of this Protocol with respect to the advance informed agreement procedure shall not apply to living modified organisms in transit.

(2) Notwithstanding Article 4 and without prejudice to any right of a Party to subject all living modified organisms to risk assessment prior to decisions on import and to set standards for contained use within its jurisdiction, the provisions of this Protocol with respect to the advance informed agreement procedure shall not apply to the transboundary movement of living modified organisms destined for contained use undertaken in accordance with the standards of the Party of import.

Article 7. Application of the Advance Informed Agreement Procedure—(1) Subject to Articles 5 and 6, the advance informed agreement procedure in Articles 8 to 10 and 12 shall apply prior to the first intentional transboundary movement of living modified organisms for intentional introduction into the environment of the Party of import.

(2) "Intentional introduction into the environment" in paragraph 1 above, does not refer to living modified organisms intended for direct use as food or feed, or for processing.

(3) Article 11 shall apply prior to the first transboundary movement of living modified organisms intended for direct use as food or feed, or for processing.

(4) The advance informed agreement procedure shall not apply to the intentional transboundary movement of living modified organisms identified in a decision of the Conference of the Parties serving as the meeting of the Parties to this Protocol as being not

likely to have adverse effects on the conservation and sustainable use of biological diversity, taking also into account risks to human health.

Article 8. Notification—(1) The Party of export shall notify, or require the exporter to ensure notification to, in writing, the competent national authority of the Party of import prior to the intentional transboundary movement of a living modified organism that falls within the scope of Article 7, paragraph 1. The notification shall contain, at a minimum, the information specified in Annex I.

(2) The Party of export shall ensure that there is a legal requirement for the accuracy of information provided by the exporter.

Article 9. Acknowledgement of Receipt of Notification—(1) The Party of import shall acknowledge receipt of the notification, in writing, to the notifier within ninety days of its receipt.

(2) The acknowledgement shall state:

(*a*) The date of receipt of the notification;

(*b*) Whether the notification, prima facie, contains the information referred to in Article 8;

(*c*) Whether to proceed according to the domestic regulatory framework of the Party of import or according to the procedure specified in Article 10.

(3) The domestic regulatory framework referred to in paragraph 2 (*c*) above, shall be consistent with this Protocol.

(4) A failure by the Party of import to acknowledge receipt of a notification shall not imply its consent to an intentional transboundary movement.

Article 10. Decision Procedure—(1) Decisions taken by the Party of import shall be in accordance with Article 15.

(2) The Party of import shall, within the period of time referred to in Article 9, inform the notifier, in writing, whether the intentional transboundary movement may proceed:

(*a*) Only after the Party of import has given its written consent; or

(*b*) After no less than ninety days without a subsequent written consent.

(3) Within two hundred and seventy days of the date of receipt of notification, the Party of import shall communicate, in writing, to the notifier and to the Biosafety Clearing-House the decision referred to in paragraph 2 (*a*) above:

(*a*) Approving the import, with or without conditions, including how the decision will apply to subsequent imports of the same living modified organism;

(*b*) Prohibiting the import;

(*c*) Requesting additional relevant information in accordance with its domestic regulatory framework or Annex I; in calculating the time within which the Party of import is to respond, the number of days it has to wait for additional relevant information shall not be taken into account; or

(*d*) Informing the notifier that the period specified in this paragraph is extended by a defined period of time.

(4) Except in a case in which consent is unconditional, a decision under paragraph 3 above, shall set out the reasons on which it is based.

(5) A failure by the Party of import to communicate its decision within two hundred and seventy days of the date of receipt of the notification shall not imply its consent to an intentional transboundary movement.

(6) Lack of scientific certainty due to insufficient relevant scientific information and knowledge regarding the extent of the potential adverse effects of a living modified organism on the conservation and sustainable use of biological diversity in the Party of import, taking also into account risks to human health, shall not prevent that Party from taking a decision, as appropriate, with regard to the import of the living modified organism in question as referred to in paragraph 3 above, in order to avoid or minimize such potential adverse effects.

(7) The Conference of the Parties serving as the meeting of the Parties shall, at its first meeting, decide upon appropriate procedures and mechanisms to facilitate decision-making by Parties of import.

Article 11. Procedure for Living Modified Organisms Intended for Direct use as Food or Feed, or for Processing—(1) A Party that makes a final decision regarding domestic use, including placing on the market, of a living modified organism that may be subject to transboundary movement for direct use as food or feed, or for processing shall, within fifteen days of making that decision, inform the Parties through the Biosafety Clearing-House. This information shall contain, at a minimum, the information specified in Annex II. The Party shall provide a copy of the information, in writing, to the national focal point of each Party that informs the Secretariat in advance that it does not have access to the Biosafety Clearing-House. This provision shall not apply to decisions regarding field trials.

(2) The Party making a decision under paragraph 1 above, shall ensure that there is a legal requirement for the accuracy of information provided by the applicant.

(3) Any Party may request additional information from the authority identified in paragraph (*b*) of Annex II.

(4) A Party may take a decision on the import of living modified organisms intended for direct use as food or feed, or for processing, under its domestic regulatory framework that is consistent with the objective of this Protocol.

(5) Each Party shall make available to the Biosafety Clearing-House copies of any national laws, regulations and guidelines applicable to the import of living modified organisms intended for direct use as food or feed, or for processing, if available.

(6) A developing country Party or a Party with an economy in transition may, in the absence of the domestic regulatory framework referred to in paragraph 4 above, and in exercise of its domestic jurisdiction, declare through the Biosafety Clearing-House that its decision prior to the first import of a living modified organism intended for direct use as food or feed, or for processing, on which information has been provided under paragraph 1 above, will be taken according to the following:

(*a*) A risk assessment undertaken in accordance with Annex III; and

(*b*) A decision made within a predictable timeframe, not exceeding two hundred and seventy days.

(7) Failure by a Party to communicate its decision according to paragraph 6 above, shall not imply its consent or refusal to the import of a living modified organism intended for direct use as food or feed, or for processing, unless otherwise specified by the Party.

(8) Lack of scientific certainty due to insufficient relevant scientific information and knowledge regarding the extent of the potential adverse effects of a living modified organism on the conservation and sustainable use of biological diversity in the Party of import, taking also into account risks to human health, shall not prevent that Party from taking a decision, as appropriate, with

regard to the import of that living modified organism intended for direct use as food or feed, or for processing, in order to avoid or minimize such potential adverse effects.

(9) A Party may indicate its needs for financial and technical assistance and capacity-building with respect to living modified organisms intended for direct use as food or feed, or for processing. Parties shall cooperate to meet these needs in accordance with Articles 22 and 28.

Article 12. Review of Decisions—(1) A Party of import may, at any time, in light of new scientific information on potential adverse effects on the conservation and sustainable use of biological diversity, taking also into account the risks to human health, review and change a decision regarding an intentional transboundary movement. In such case, the Party shall, within thirty days, inform any notifier that has previously notified movements of the living modified organism referred to in such decision, as well as the Biosafety Clearing-House, and shall set out the reasons for its decision.

(2) A Party of export or a notifier may request the Party of import to review a decision it has made in respect of it under Article 10 where the Party of export or the notifier considers that:

(*a*) A change in circumstances has occurred that may influence the outcome of the risk assessment upon which the decision was based; or

(*b*) Additional relevant scientific or technical information has become available.

(3) The Party of import shall respond in writing to such a request within ninety days and set out the reasons for its decision.

(4) The Party of import may, at its discretion, require a risk assessment for subsequent imports.

Article 13. Simplified Procedure—(1) A Party of import may, provided that adequate measures are applied to ensure the safe intentional transboundary movement of living modified organisms in accordance with the objective of this Protocol, specify in advance to the Biosafety Clearing-House:

(*a*) Cases in which intentional transboundary movement to it may take place at the same time as the movement is notified to the Party of import; and

(*b*) Imports of living modified organisms to it to be exempted from the advance informed agreement procedure.

Notifications under subparagraph (*a*) above, may apply to subsequent similar movements to the same Party.

(2) The information relating to an intentional transboundary movement that is to be provided in the notifications referred to in paragraph 1 (*a*) above, shall be the information specified in Annex I.

Article 14. Bilateral, Regional and Multilateral Agreements and Arrangements—(1) Parties may enter into bilateral, regional and multilateral agreements and arrangements regarding intentional transboundary movements of living modified organisms, consistent with the objective of this Protocol and provided that such agreements and arrangements do not result in a lower level of protection than that provided for by the Protocol.

(2) The Parties shall inform each other, through the Biosafety Clearing-House, of any such bilateral, regional and multilateral agreements and arrangements that they have entered into before or after the date of entry into force of this Protocol.

(3) The provisions of this Protocol shall not affect intentional transboundary movements that take place pursuant to such agreements and arrangements as between the parties to those agreements or arrangements.

(4) Any Party may determine that its domestic regulations shall apply with respect to specific imports to it and shall notify the Biosafety Clearing-House of its decision.

Article 15. Risk Assessment—(1) Risk assessments undertaken pursuant to this Protocol shall be carried out in a scientifically sound manner, in accordance with Annex III and taking into account recognized risk assessment techniques. Such risk assessments shall be based, at a minimum, on information provided in accordance with Article 8 and other available scientific evidence in order to identify and evaluate the possible adverse effects of living modified organisms on the conservation and sustainable use of biological diversity, taking also into account risks to human health.

(2) The Party of import shall ensure that risk assessments are carried out for decisions taken under Article 10. It may require the exporter to carry out the risk assessment.

(3) The cost of risk assessment shall be borne by the notifier if the Party of import so requires.

Article 16. Risk Management—(1) The Parties shall, taking into account Article 8 (*g*) of the Convention, establish and maintain appropriate mechanisms, measures and strategies to regulate manage and control risks identified in the risk assessment provisions of this Protocol associated with the use, handling and transboundary movement of living modified organisms.

(2) Measures based on risk assessment shall be imposed to the extent necessary to prevent adverse effects of the living modified organism on the conservation and sustainable use of biological diversity, taking also into account risks to human health, within the territory of the Party of import.

(3) Each Party shall take appropriate measures to prevent unintentional transboundary movements of living modified organisms, including such measures as requiring a risk assessment to be carried out prior to the first release of a living modified organism.

(4) Without prejudice to paragraph 2 above, each Party shall endeavour to ensure that any living modified organism, whether imported or locally developed, has undergone an appropriate period of observation that is commensurate with its life-cycle or generation time before it is put to its intended use.

(5) Parties shall cooperate with a view to:

(*a*) Identifying living modified organisms or specific traits of living modified organisms that may have adverse effects on the conservation and sustainable use of biological diversity, taking also into account risks to human health; and

(*b*) Taking appropriate measures regarding the treatment of such living modified organisms or specific traits.

Article 17. Unintentional Transboundary Movements and Emergency Measures—(1) Each Party shall take appropriate measures to notify affected or potentially affected States, the Biosafety Clearing-House and, where appropriate, relevant international organizations, when it knows of an occurrence under its jurisdiction resulting in a release that leads, or may lead, to an unintentional transboundary movement of a living modified organism that is likely to have significant adverse effects on the conservation and sustainable use of biological diversity, taking also into account risks to human health in such States. The notification shall be provided as soon as the Party knows of the above situation.

(2) Each Party shall, no later than the date of entry into force of this Protocol for it, make available to the Biosafety Clearing-House

the relevant details setting out its point of contact for the purposes of receiving notifications under this Article.

(3) Any notification arising from paragraph 1 above, should include:

(*a*) Available relevant information on the estimated quantities and relevant characteristics and/or traits of the living modified organism;

(*b*) Information on the circumstances and estimated date of the release, and on the use of the living modified organism in the originating Party;

(*c*) Any available information about the possible adverse effects on the conservation and sustainable use of biological diversity, taking also into account risks to human health, as well as available information about possible risk management measures;

(*d*) Any other relevant information; and

(*e*) A point of contact for further information.

(4) In order to minimize any significant adverse effects on the conservation and sustainable use of biological diversity, taking also into account risks to human health, each Party, under whose jurisdiction the release of the living modified organism referred to in paragraph 1 above, occurs, shall immediately consult the affected or potentially affected States to enable them to determine appropriate responses and initiate necessary action, including emergency measures.

Article 18. Handling, Transport, Packaging and Identification—(1) In order to avoid adverse effects on the conservation and sustainable use of biological diversity, taking also into account risks to human health, each Party shall take necessary measures to require that living modified organisms that are subject to intentional transboundary movement within the scope of this Protocol are handled, packaged and transported under conditions of safety, taking into consideration relevant international rules and standards.

(2) Each Party shall take measures to require that documentation accompanying:

(*a*) Living modified organisms that are intended for direct use as food or feed, or for processing, clearly identifies that they "may contain" living modified organisms and are not intended for intentional introduction into the environment, as well as a contact point for further information. The

Conference of the Parties serving as the meeting of the Parties to this Protocol shall take a decision on the detailed requirements for this purpose, including specification of their identity and any unique identification, no later than two years after the date of entry into force of this Protocol;

(*b*) Living modified organisms that are destined for contained use clearly identifies them as living modified organisms; and specifies any requirements for the safe handling, storage, transport and use, the contact point for further information, including the name and address of the individual and institution to whom the living modified organisms are consigned; and

(*c*) Living modified organisms that are intended for intentional introduction into the environment of the Party of import and any other living modified organisms within the scope of the Protocol, clearly identifies them as living modified organisms; specifies the identity and relevant traits and/or characteristics, any requirements for the safe handling, storage, transport and use, the contact point for further information and, as appropriate, the name and address of the importer and exporter; and contains a declaration that the movement is in conformity with the requirements of this Protocol applicable to the exporter.

(3) The Conference of the Parties serving as the meeting of the Parties to this Protocol shall consider the need for and modalities of developing standards with regard to identification, handling, packaging and transport practices, in consultation with other relevant international bodies.

Article 19. Competent National Authorities and National Focal Points—(1) Each Party shall designate one national focal point to be responsible on its behalf for liaison with the Secretariat. Each Party shall also designate one or more competent national authorities, which shall be responsible for performing the administrative functions required by this Protocol and which shall be authorized to act on its behalf with respect to those functions. A Party may designate a single entity to fulfil the functions of both focal point and competent national authority.

(2) Each Party shall, no later than the date of entry into force of this Protocol for it, notify the Secretariat of the names and addresses of its focal point and its competent national authority or authorities.

Where a Party designates more than one competent national authority, it shall convey to the Secretariat, with its notification thereof, relevant information on the respective responsibilities of those authorities. Where applicable, such information shall, at a minimum, specify which competent authority is responsible for which type of living modified organism. Each Party shall forthwith notify the Secretariat of any changes in the designation of its national focal point or in the name and address or responsibilities of its competent national authority or authorities.

(3) The Secretariat shall forthwith inform the Parties of the notifications it receives under paragraph 2 above, and shall also make such information available through the Biosafety Clearing-House.

Article 20. Information Sharing and the Biosafety Clearing-House—(1) A Biosafety Clearing-House is hereby established as part of the clearing-house mechanism under Article 18, paragraph 3, of the Convention, in order to:

(*a*) Facilitate the exchange of scientific, technical, environmental and legal information on, and experience with, living modified organisms; and

(*b*) Assist Parties to implement the Protocol, taking into account the special needs of developing country Parties, in particular the least developed and small island developing States among them, and countries with economies in transition as well as countries that are centres of origin and centres of genetic diversity.

(2) The Biosafety Clearing-House shall serve as a means through which information is made available for the purposes of paragraph 1 above. It shall provide access to information made available by the Parties relevant to the implementation of the Protocol. It shall also provide access, where possible, to other international biosafety information exchange mechanisms.

(3) Without prejudice to the protection of confidential information, each Party shall make available to the Biosafety Clearing-House any information required to be made available to the Biosafety Clearing-House under this Protocol, and:

(*a*) Any existing laws, regulations and guidelines for implementation of the Protocol, as well as information required by the Parties for the advance informed agreement procedure;

(*b*) Any bilateral, regional and multilateral agreements and arrangements;

(*c*) Summaries of its risk assessments or environmental reviews of living modified organisms generated by its regulatory process, and carried out in accordance with Article 15, including, where appropriate, relevant information regarding products thereof, namely, processed materials that are of living modified organism origin, containing detectable novel combinations of replicable genetic material obtained through the use of modern biotechnology;

(*d*) Its final decisions regarding the importation or release of living modified organisms; and

(*e*) Reports submitted by it pursuant to Article 33, including those on implementation of the advance informed agreement procedure.

(4) The modalities of the operation of the Biosafety Clearing-House, including reports on its activities, shall be considered and decided upon by the Conference of the Parties serving as the meeting of the Parties to this Protocol at its first meeting, and kept under review thereafter.

Article 21. Confidential Information—(1) The Party of import shall permit the notifier to identify information submitted under the procedures of this Protocol or required by the Party of import as part of the advance informed agreement procedure of the Protocol that is to be treated as confidential. Justification shall be given in such cases upon request.

(2) The Party of import shall consult the notifier if it decides that information identified by the notifier as confidential does not qualify for such treatment and shall, prior to any disclosure, inform the notifier of its decision, providing reasons on request, as well as an opportunity for consultation and for an internal review of the decision prior to disclosure.

(3) Each Party shall protect confidential information received under this Protocol, including any confidential information received in the context of the advance informed agreement procedure of the Protocol. Each Party shall ensure that it has procedures to protect such information and shall protect the confidentiality of such information in a manner no less favourable than its treatment of confidential information in connection with domestically produced living modified organisms.

(4) The Party of import shall not use such information for a commercial purpose, except with the written consent of the notifier.

(5) If a notifier withdraws or has withdrawn a notification, the Party of import shall respect the confidentiality of commercial and industrial information, including research and development information as well as information on which the Party and the notifier disagree as to its confidentiality.

(6) Without prejudice to paragraph 5 above, the following information shall not be considered confidential:

(*a*) The name and address of the notifier;

(*b*) A general description of the living modified organism or organisms;

(*c*) A summary of the risk assessment of the effects on the conservation and sustainable use of biological diversity, taking also into account risks to human health; and

(*d*) Any methods and plans for emergency response.

Article 22. Capacity-Building—(1) The Parties shall cooperate in the development and/or strengthening of human resources and institutional capacities in biosafety, including biotechnology to the extent that it is required for biosafety, for the purpose of the effective implementation of this Protocol, in developing country Parties, in particular the least developed and small island developing States among them, and in Parties with economies in transition, including through existing global, regional, subregional and national institutions and organizations and, as appropriate, through facilitating private sector involvement.

(2) For the purposes of implementing paragraph 1 above, in relation to cooperation, the needs of developing country Parties, in particular the least developed and small island developing States among them, for financial resources and access to and transfer of technology and know-how in accordance with the relevant provisions of the Convention, shall be taken fully into account for capacity-building in biosafety. Cooperation in capacity-building shall, subject to the different situation, capabilities and requirements of each Party, include scientific and technical training in the proper and safe management of biotechnology, and in the use of risk assessment and risk management for biosafety, and the enhancement of technological and institutional capacities in biosafety. The needs of Parties with economies in transition shall also be taken fully into account for such capacity-building in biosafety.

Article 23. Public Awareness and Participation—(1) The Parties shall:

(*a*) Promote and facilitate public awareness, education and participation concerning the safe transfer, handling and use of living modified organisms in relation to the conservation and sustainable use of biological diversity, taking also into account risks to human health. In doing so, the Parties shall cooperate, as appropriate, with other States and international bodies;

(*b*) Endeavour to ensure that public awareness and education encompass access to information on living modified organisms identified in accordance with this Protocol that may be imported.

(2) The Parties shall, in accordance with their respective laws and regulations, consult the public in the decision-making process regarding living modified organisms and shall make the results of such decisions available to the public, while respecting confidential information in accordance with Article 21.

(3) Each Party shall endeavour to inform its public about the means of public access to the Biosafety Clearing-House.

Article 24. Non-Parties—(1) Transboundary movements of living modified organisms between Parties and non-Parties shall be consistent with the objective of this Protocol. The Parties may enter into bilateral, regional and multilateral agreements and arrangements with non-Parties regarding such transboundary movements.

(2) The Parties shall encourage non-Parties to adhere to this Protocol and to contribute appropriate information to the Biosafety Clearing-House on living modified organisms released in, or moved into or out of, areas within their national jurisdictions.

Article 25. Illegal Transboundary Movements—(1) Each Party shall adopt appropriate domestic measures aimed at preventing and, if appropriate, penalizing transboundary movements of living modified organisms carried out in contravention of its domestic measures to implement this Protocol. Such movements shall be deemed illegal transboundary movements.

(2) In the case of an illegal transboundary movement, the affected Party may request the Party of origin to dispose, at its own expense, of the living modified organism in question by repatriation or destruction, as appropriate.

(3) Each Party shall make available to the Biosafety Clearing-House information concerning cases of illegal transboundary movements pertaining to it.

Article 26. Socio-Economic Considerations—(1) The Parties, in reaching a decision on import under this Protocol or under its domestic measures implementing the Protocol, may take into account, consistent with their international obligations, socio-economic considerations arising from the impact of living modified organisms on the conservation and sustainable use of biological diversity, especially with regard to the value of biological diversity to indigenous and local communities.

(2) The Parties are encouraged to cooperate on research and information exchange on any socio-economic impacts of living modified organisms, especially on indigenous and local communities.

Article 27. Liability and Redress—The Conference of the Parties serving as the meeting of the Parties to this Protocol shall, at its first meeting, adopt a process with respect to the appropriate elaboration of international rules and procedures in the field of liability and redress for damage resulting from transboundary movements of living modified organisms, analysing and taking due account of the ongoing processes in international law on these matters, and shall endeavour to complete this process within four years.

Article 28. Financial Mechanism and Resources—(1) In considering financial resources for the implementation of this Protocol, the Parties shall take into account the provisions of Article 20 of the Convention.

(2) The financial mechanism established in Article 21 of the Convention shall, through the institutional structure entrusted with its operation, be the financial mechanism for this Protocol.

(3) Regarding the capacity-building referred to in Article 22 of this Protocol, the Conference of the Parties serving as the meeting of the Parties to this Protocol, in providing guidance with respect to the financial mechanism referred to in paragraph 2 above, for consideration by the Conference of the Parties, shall take into account the need for financial resources by developing country Parties, in particular the least developed and the small island developing States among them.

(4) In the context of paragraph 1 above, the Parties shall also take into account the needs of the developing country Parties, in

particular the least developed and the small island developing States among them, and of the Parties with economies in transition, in their efforts to identify and implement their capacity-building requirements for the purposes of the implementation of this Protocol.

(5) The guidance to the financial mechanism of the Convention in relevant decisions of the Conference of the Parties, including those agreed before the adoption of this Protocol, shall apply, *mutatis mutandis*, to the provisions of this Article.

(6) The developed country Parties may also provide, and the developing country Parties and the Parties with economies in transition avail themselves of, financial and technological resources for the implementation of the provisions of this Protocol through bilateral, regional and multilateral channels.

Article 29. Conference of the Parties Serving as the Meeting of the Parties to this Protocol—(1) The Conference of the Parties shall serve as the meeting of the Parties to this Protocol.

(2) Parties to the Convention that are not Parties to this Protocol may participate as observers in the proceedings of any meeting of the Conference of the Parties serving as the meeting of the Parties to this Protocol. When the Conference of the Parties serves as the meeting of the Parties to this Protocol, decisions under this Protocol shall be taken only by those that are Parties to it.

(3) When the Conference of the Parties serves as the meeting of the Parties to this Protocol, any member of the bureau of the Conference of the Parties representing a Party to the Convention but, at that time, not a Party to this Protocol, shall be substituted by a member to be elected by and from among the Parties to this Protocol.

(4) The Conference of the Parties serving as the meeting of the Parties to this Protocol shall keep under regular review the implementation of this Protocol and shall make, within its mandate, the decisions necessary to promote its effective implementation. It shall perform the functions assigned to it by this Protocol and shall:

(*a*) Make recommendations on any matters necessary for the implementation of this Protocol;

(*b*) Establish such subsidiary bodies as are deemed necessary for the implementation of this Protocol;

(*c*) Seek and utilize, where appropriate, the services and cooperation of, and information provided by, competent international organizations and intergovernmental and non-governmental bodies;

(*d*) Establish the form and the intervals for transmitting the information to be submitted in accordance with Article 33 of this Protocol and consider such information as well as reports submitted by any subsidiary body;

(*e*) Consider and adopt, as required, amendments to this Protocol and its annexes, as well as any additional annexes to this Protocol, that are deemed necessary for the implementation of this Protocol; and

(*f*) Exercise such other functions as may be required for the implementation of this Protocol.

(5) The rules of procedure of the Conference of the Parties and financial rules of the Convention shall be applied, *mutatis mutandis,* under this Protocol, except as may be otherwise decided by consensus by the Conference of the Parties serving as the meeting of the Parties to this Protocol.

(6) The first meeting of the Conference of the Parties serving as the meeting of the Parties to this Protocol shall be convened by the Secretariat in conjunction with the first meeting of the Conference of the Parties that is scheduled after the date of the entry into force of this Protocol. Subsequent ordinary meetings of the Conference of the Parties serving as the meeting of the Parties to this Protocol shall be held in conjunction with ordinary meetings of the Conference of the Parties, unless otherwise decided by the Conference of the Parties serving as the meeting of the Parties to this Protocol.

(7) Extraordinary meetings of the Conference of the Parties serving as the meeting of the Parties to this Protocol shall be held at such other times as may be deemed necessary by the Conference of the Parties serving as the meeting of the Parties to this Protocol, or at the written request of any Party, provided that, within six months of the request being communicated to the Parties by the Secretariat, it is supported by at least one third of the Parties.

(8) The United Nations, its specialized agencies and the International Atomic Energy Agency, as well as any State member thereof or observers thereto not party to the Convention, may be represented as observers at meetings of the Conference of the Parties serving as the meeting of the Parties to this Protocol. Any body or agency, whether national or international, governmental or non-governmental, that is qualified in matters covered by this Protocol

and that has informed the Secretariat of its wish to be represented at a meeting of the Conference of the Parties serving as a meeting of the Parties to this Protocol as an observer, may be so admitted, unless at least one third of the Parties present object. Except as otherwise provided in this Article, the admission and participation of observers shall be subject to the rules of procedure, as referred to in paragraph 5 above.

Article 30. Subsidiary Bodies—(1) Any subsidiary body established by or under the Convention may, upon a decision by the Conference of the Parties serving as the meeting of the Parties to this Protocol, serve the Protocol, in which case the meeting of the Parties shall specify which functions that body shall exercise.

(2) Parties to the Convention that are not Parties to this Protocol may participate as observers in the proceedings of any meeting of any such subsidiary bodies. When a subsidiary body of the Convention serves as a subsidiary body to this Protocol, decisions under the Protocol shall be taken only by the Parties to the Protocol.

(3) When a subsidiary body of the Convention exercises its functions with regard to matters concerning this Protocol, any member of the bureau of that subsidiary body representing a Party to the Convention but, at that time, not a Party to the Protocol, shall be substituted by a member to be elected by and from among the Parties to the Protocol.

Article 31. Secretariat—(1) The Secretariat established by Article 24 of the Convention shall serve as the secretariat to this Protocol.

(2) Article 24, paragraph 1, of the Convention on the functions of the Secretariat shall apply, *mutatis mutandis*, to this Protocol.

(3) To the extent that they are distinct, the costs of the secretariat services for this Protocol shall be met by the Parties hereto. The Conference of the Parties serving as the meeting of the Parties to this Protocol shall, at its first meeting, decide on the necessary budgetary arrangements to this end.

Article 32. Relationship with the Convention—Except as otherwise provided in this Protocol, the provisions of the Convention relating to its protocols shall apply to this Protocol.

Article 33. Monitoring and Reporting—Each Party shall monitor the implementation of its obligations under this Protocol, and shall, at intervals to be determined by the Conference of the Parties serving as the meeting of the Parties to this Protocol, report to the Conference of the Parties serving as the meeting of the Parties to this Protocol on measures that it has taken to implement the Protocol.

Article 34. Compliance—The Conference of the Parties serving as the meeting of the Parties to this Protocol shall, at its first meeting, consider and approve cooperative procedures and institutional mechanisms to promote compliance with the provisions of this Protocol and to address cases of non-compliance. These procedures and mechanisms shall include provisions to offer advice or assistance, where appropriate. They shall be separate from, and without prejudice to, the dispute settlement procedures and mechanisms established by Article 27 of the Convention.

Article 35. Assessment and Review—The Conference of the Parties serving as the meeting of the Parties to this Protocol shall undertake, five years after the entry into force of this Protocol and at least every five years thereafter, an evaluation of the effectiveness of the Protocol, including an assessment of its procedures and annexes.

Article 36. Signature—This Protocol shall be open for signature at the United Nations Office at Nairobi by States and regional economic integration organizations from 15 to 26 May 2000, and at United Nations Headquarters in New York from 5 June 2000 to 4 June 2001.

Article 37. Entry Into Force—(1) This Protocol shall enter into force on the ninetieth day after the date of deposit of the fiftieth instrument of ratification, acceptance, approval or accession by States or regional economic integration organizations that are Parties to the Convention.

(2) This Protocol shall enter into force for a State or regional economic integration organization that ratifies, accepts or approves this Protocol or accedes thereto after its entry into force pursuant to paragraph 1 above, on the ninetieth day after the date on which that State or regional economic integration organization deposits its instrument of ratification, acceptance, approval or accession, or on the date on which the Convention enters into force for that State or regional economic integration organization, whichever shall be the later.

(3) For the purposes of paragraphs 1 and 2 above, any instrument deposited by a regional economic integration organization shall not be counted as additional to those deposited by member States of such organization.

Article 38. Reservations—No reservations may be made to this Protocol.

Article 39. Withdrawal—(1) At any time after two years from the date on which this Protocol has entered into force for a Party, that Party may withdraw from the Protocol by giving written notification to the Depositary.

(2) Any such withdrawal shall take place upon expiry of one year after the date of its receipt by the Depositary, or on such later date as may be specified in the notification of the withdrawal.

Article 40. Authentic Texts—The original of this Protocol, of which the Arabic, Chinese, English, French, Russian and Spanish texts are equally authentic, shall be deposited with the Secretary-General of the United Nations.

In witness whereof the undersigned, being duly authorized to that effect, have signed this Protocol.

Done at Montreal on this twenty-ninth day of January, two thousand.

Annex I
Information Required in Notifications under Articles 8, 10 and 13

(*a*) Name, address and contact details of the exporter.

(*b*) Name, address and contact details of the importer.

(*c*) Name and identity of the living modified organism, as well as the domestic classification, if any, of the biosafety level of the living modified organism in the State of export.

(*d*) Intended date or dates of the transboundary movement, if known.

(*e*) Taxonomic status, common name, point of collection or acquisition, and characteristics of recipient organism or parental organisms related to biosafety.

(*f*) Centres of origin and centres of genetic diversity, if known, of the recipient organism and/or the parental organisms and a description of the habitats where the organisms may persist or proliferate.

(*g*) Taxonomic status, common name, point of collection or acquisition, and characteristics of the donor organism or organisms related to biosafety.

(*h*) Description of the nucleic acid or the modification introduced, the technique used, and the resulting characteristics of the living modified organism.

(*i*) Intended use of the living modified organism or products thereof, namely, processed materials that are of living modified organism origin, containing detectable novel combinations of replicable genetic material obtained through the use of modern biotechnology.

(*j*) Quantity or volume of the living modified organism to be transferred.

(*k*) A previous and existing risk assessment report consistent with Annex III.

(*l*) Suggested methods for the safe handling, storage, transport and use, including packaging, labelling, documentation, disposal and contingency procedures, where appropriate.

(*m*) Regulatory status of the living modified organism within the State of export (for example, whether it is prohibited in the State of export, whether there are other restrictions, or whether it has been approved for general release) and, if the living modified organism is banned in the State of export, the reason or reasons for the ban.

(*n*) Result and purpose of any notification by the exporter to other States regarding the living modified organism to be transferred.

(*o*) A declaration that the above-mentioned information is factually correct.

Annex II
Information Required Concerning Living Modified Organisms Intended for Direct use as Food or Feed, or for Processing Under Article 11

(*a*) The name and contact details of the applicant for a decision for domestic use.

(*b*) The name and contact details of the authority responsible for the decision.

(*c*) Name and identity of the living modified organism.

(*d*) Description of the gene modification, the technique used, and the resulting characteristics of the living modified organism.

(*e*) Any unique identification of the living modified organism.
(*f*) Taxonomic status, common name, point of collection or acquisition, and characteristics of recipient organism or parental organisms related to biosafety.
(*g*) Centres of origin and centres of genetic diversity, if known, of the recipient organism and/or the parental organisms and a description of the habitats where the organisms may persist or proliferate.
(*h*) Taxonomic status, common name, point of collection or acquisition, and characteristics of the donor organism or organisms related to biosafety.
(*i*) Approved uses of the living modified organism.
(*j*) A risk assessment report consistent with Annex III.
(*k*) Suggested methods for the safe handling, storage, transport and use, including packaging, labelling, documentation, disposal and contingency procedures, where appropriate.

Annex III
Risk Assessment

Objective

The objective of risk assessment, under this Protocol, is to identify and evaluate the potential adverse effects of living modified organisms on the conservation and sustainable use of biological diversity in the likely potential receiving environment, taking also into account risks to human health.

Use of Risk Assessment

Risk assessment is, *inter alia*, used by competent authorities to make informed decisions regarding living modified organisms.

General Principles

Risk assessment should be carried out in a scientifically sound and transparent manner, and can take into account expert advice of, and guidelines developed by, relevant international organizations.

Lack of scientific knowledge or scientific consensus should not necessarily be interpreted as indicating a particular level of risk, an absence of risk, or an acceptable risk.

Risks associated with living modified organisms or products thereof, namely, processed materials that are of living modified organism origin, containing detectable novel combinations of replicable genetic material obtained through the use of modern biotechnology, should be considered in the context of the risks posed by the non-modified recipients or parental organisms in the likely potential receiving environment.

Risk assessment should be carried out on a case-by-case basis. The required information may vary in nature and level of detail from case to case, depending on the living modified organism concerned, its intended use and the likely potential receiving environment.

Methodology

The process of risk assessment may on the one hand give rise to a need for further information about specific subjects, which may be identified and requested during the assessment process, while on the other hand information on other subjects may not be relevant in some instances.

To fulfil its objective, risk assessment entails, as appropriate, the following steps:

(*a*) An identification of any novel genotypic and phenotypic characteristics associated with the living modified organism that may have adverse effects on biological diversity in the likely potential receiving environment, taking also into account risks to human health;

(*b*) An evaluation of the likelihood of these adverse effects being realized, taking into account the level and kind of exposure of the likely potential receiving environment to the living modified organism;

(*c*) An evaluation of the consequences should these adverse effects be realized;

(*d*) An estimation of the overall risk posed by the living modified organism based on the evaluation of the likelihood and consequences of the identified adverse effects being realized;

(*e*) A recommendation as to whether or not the risks are acceptable or manageable, including, where necessary, identification of strategies to manage these risks; and

(*f*) Where there is uncertainty regarding the level of risk, it may be addressed by requesting further information on the specific issues of concern or by implementing appropriate risk management strategies and/or monitoring the living modified organism in the receiving environment.

Points to Consider

Depending on the case, risk assessment takes into account the relevant technical and scientific details regarding the characteristics of the following subjects:

(*a*) *Recipient organism or parental organisms*. The biological characteristics of the recipient organism or parental organisms, including information on taxonomic status, common name, origin, centres of origin and centres of genetic diversity, if known, and a description of the habitat where the organisms may persist or proliferate;

(*b*) *Donor organism or organisms*. Taxonomic status and common name, source, and the relevant biological characteristics of the donor organisms;

(*c*) *Vector*. Characteristics of the vector, including its identity, if any, and its source or origin, and its host range;

(*d*) *Insert or inserts and/or characteristics of modification*. Genetic characteristics of the inserted nucleic acid and the function it specifies, and/or characteristics of the modification introduced;

(*e*) *Living modified organism*. Identity of the living modified organism, and the differences between the biological characteristics of the living modified organism and those of the recipient organism or parental organisms;

(*f*) *Detection and identification of the living modified organism*. Suggested detection and identification methods and their specificity, sensitivity and reliability;

(*g*) *Information relating to the intended use*. Information relating to the intended use of the living modified organism, including new or changed use compared to the recipient organism or parental organisms; and

(h) *Receiving environment.* Information on the location, geographical, climatic and ecological characteristics, including relevant information on biological diversity and centres of origin of the likely potential receiving environment.

"Friends of the Convention on Life on Earth" in Gambia will promote action for the planet by civil society. The Gambia becomes the first Party to the Convention on Biological Diversity to establish a forum for citizen engagement on action for the 2010 biodiversity target.

Business and industry have a crucial role to play in all aspects of sustainable development. Statement to business and industry by Ahmed Djoghlaf

The successful conclusion of your negotiations will put an end to the current uncertainty. Remarks by Ahmed Djoghlaf at the Opening session of the fourth meeting of the Ad Hoc Open-ended Inter-Sessional Working Group on Access and Benefit-sharing.

There will be no biological diversity on our planet if there is no cultural diversity. Remarks by Doña Cristina Narbona Ruiz, Minister of the Environment for Spain at the fourth meeting of the Ad Hoc Open-ended Inter-Sessional Working Group on Article 8(*j*) and Related Provisions. (Text in Spanish)

Nature is at the heart of the cultures and values of traditional societies. Remarks by Executive Secretary Ahmed Djoghlaf, at the opening of the fourth meeting of the Ad Hoc Open-ended Inter-Sessional Working Group on Article 8(*j*) and Related Provisions.

The Virtual Curitiba Biodiversity Conference. At a press conference with the Minister of the Environment for Brazil, H.E.M. Minister Marina Silva, on the preparation of the International Ministerial Curitiba Biodiversity Conference to be held in March, Dr. Djoghlaf announced the official launch of the Virtual Curitiba Biodiversity Conference.

Message of Ahmed Djoghlaf to the citizens of Brazil. I call on all citizens of Brazil to ensure that the Curitiba International Conference be remembered by future generations as the birth place of a new era of enhanced implementation of the Convention.

Indigenous and local communities have important role to play in implementation of Biodiversity Convention: Ahmed Djoghlaf. Montreal–10 January 2006. Stressing the role of traditional knowledge and the practices of indigenous and local communities in biodiversity conservation, Ahmed Djoghlaf, the Executive Secretary to the Convention on Biological Diversity, reiterated the importance of the participation of indigenous and local communities in the Convention process.

World community urged to protect biodiversity in deserts-theme for International Day for Biological Diversity announced. "One in six people depends on these fragile ecosystems for their survival," said Mr. Djoghlaf, Executive Secretary to the Convention. "These are the poorest of the poor and any further degradation of the biodiversity of drylands will lead to increased poverty."

Ahmed Djoghlaf assumes duties as Executive Secretary to the Convention on Biological Diversity: Calls for a global alliance to combat biodiversity loss. Montreal, 3 January 2006—Calling on the citizens of the world to join in the effort to preserve life on earth, a 52 year-old Algerian man with a passion for poetry and forest walks takes up the helm of the world's key biological treaty today. United Nations Secretary-General Kofi Annan, appointed Mr. Ahmed Djoghlaf, as the new Executive Secretary to the Convention on Biological Diversity (CBD).

Chapter 4

Global Law, Policy, Action Plan to Combat Desertification

UNITED NATIONS CONVENTION TO COMBAT DESERTIFICATION (UNCCD) IN COUNTRIES EXPERIENCING SERIOUS DROUGHT AND/OR DESERTIFICATION, PARTICULARLY IN AFRICA

The Parties to this Convention,

Affirming that human beings in affected or threatened areas are at the centre of concerns to combat desertification and mitigate the effects of drought,

Reflecting the urgent concern of the international community, including States and international organizations, about the adverse impacts of desertification and drought,

Aware that arid, semi-arid and dry sub-humid areas together account for a significant proportion of the Earth's land area and are the habitat and source of livelihood for a large segment of its population,

Acknowledging that desertification and drought are problems of global dimension in that they affect all regions of the world and that joint action of the international community is needed to combat desertification and/or mitigate the effects of drought,

Noting the high concentration of developing countries, notably the least developed countries, among those experiencing serious drought and/or desertification, and the particularly tragic consequences of these phenomena in Africa,

Noting also that desertification is caused by complex interactions among physical, biological, political, social, cultural and economic factors,

Considering the impact of trade and relevant aspects of international economic relations on the ability of affected countries to combat desertification adequately,

Conscious that sustainable economic growth, social development and poverty eradication are priorities of affected developing countries, particularly in Africa, and are essential to meeting sustainability objectives,

Mindful that desertification and drought affect sustainable development through their interrelationships with important social problems such as poverty, poor health and nutrition, lack of food security, and those arising from migration, displacement of persons and demographic dynamics,

Appreciating the significance of the past efforts and experience of States and international organizations in combating desertification and mitigating the effects of drought, particularly in implementing the Plan of Action to Combat Desertification which was adopted at the United Nations Conference on Desertification in 1977,

Realizing that, despite efforts in the past, progress in combating desertification and mitigating the effects of drought has not met expectations and that a new and more effective approach is needed at all levels within the framework of sustainable development,

Recognizing the validity and relevance of decisions adopted at the United Nations Conference on Environment and Development, particularly of Agenda 21 and its chapter 12, which provide a basis for combating desertification,

Reaffirming in this light the commitments of developed countries as contained in paragraph 13 of chapter 33 of Agenda 21,

Recalling General Assembly resolution 47/188, particularly the priority in it prescribed for Africa, and all other relevant United Nations resolutions, decisions and programmes on desertification and drought, as well as relevant declarations by African countries and those from other regions,

Reaffirming the Rio Declaration on Environment and Development which states, in its Principle 2, that States have, in accordance with the Charter of the United Nations and the principles of international law, the sovereign right to exploit their own resources pursuant to their own environmental and developmental policies, and the responsibility to ensure that activities within their jurisdiction or control do not cause damage to the environment of other States or of areas beyond the limits of national jurisdiction,

Recognizing that national Governments play a critical role in combating desertification and mitigating the effects of drought and that progress in that respect depends on local implementation of action programmes in affected areas,

Recognizing also the importance and necessity of international cooperation and partnership in combating desertification and mitigating the effects of drought,

Recognizing further the importance of the provision to affected developing countries, particularly in Africa, of effective means, *inter alia* substantial financial resources, including new and additional funding, and access to technology, without which it will be difficult for them to implement fully their commitments under this Convention,

Expressing concern over the impact of desertification and drought on affected countries in Central Asia and the Transcaucasus,

Stressing the important role played by women in regions affected by desertification and/or drought, particularly in rural areas of developing countries, and the importance of ensuring the full participation of both men and women at all levels in programmes to combat desertification and mitigate the effects of drought,

Emphasizing the special role of non-governmental organizations and other major groups in programmes to combat desertification and mitigate the effects of drought,

Bearing in mind the relationship between desertification and other environmental problems of global dimension facing the international and national communities,

Bearing also in mind the contribution that combating desertification can make to achieving the objectives of the United Nations Framework Convention on Climate Change, the Convention on Biological Diversity and other related environmental conventions,

Believing that strategies to combat desertification and mitigate the effects of drought will be most effective if they are based on sound systematic observation and rigorous scientific knowledge and if they are continuously re-evaluated,

Recognizing the urgent need to improve the effectiveness and coordination of international cooperation to facilitate the implementation of national plans and priorities,

Determined to take appropriate action in combating desertification and mitigating the effects of drought for the benefit of present and future generations,

Have agreed as follows:

PART I
INTRODUCTION

Article 1. Use of terms—For the purposes of this Convention:

(*a*) "desertification" means land degradation in arid, semi-arid and dry sub-humid areas resulting from various factors, including climatic variations and human activities;

(*b*) "combating desertification" includes activities which are part of the integrated development of land in arid, semi-arid and dry sub-humid areas for sustainable development which are aimed at:

(*i*) prevention and/or reduction of land degradation;

(*ii*) rehabilitation of partly degraded land; and

(*iii*) reclamation of desertified land;

(*c*) "drought" means the naturally occurring phenomenon that exists when precipitation has been significantly below normal recorded levels, causing serious hydrological imbalances that adversely affect land resource production systems;

(*d*) "mitigating the effects of drought" means activities related to the prediction of drought and intended to reduce the vulnerability of society and natural systems to drought as it relates to combating desertification;

(*e*) "land" means the terrestrial bio-productive system that comprises soil, vegetation, other biota, and the ecological and hydrological processes that operate within the system;

(*f*) "land degradation" means reduction or loss, in arid, semi-arid and dry sub-humid areas, of the biological or economic productivity and complexity of rainfed cropland, irrigated cropland, or range, pasture, forest and woodlands resulting from land uses or from a process or combination of processes, including processes arising from human activities and habitation patterns, such as:

(*i*) soil erosion caused by wind and/or water;

(*ii*) deterioration of the physical, chemical and biological or economic properties of soil; and

(*iii*) long-term loss of natural vegetation;

(*g*) "arid, semi-arid and dry sub-humid areas" means areas, other than polar and sub-polar regions, in which the ratio of annual precipitation to potential evapotranspiration falls within the range from 0.05 to 0.65;

(*h*) "affected areas" means arid, semi-arid and/or dry sub-humid areas affected or threatened by desertification;

(*i*) "affected countries" means countries whose lands include, in whole or in part, affected areas;

(*j*) "regional economic integration organization" means an organization constituted by sovereign States of a given region which has competence in respect of matters governed by this Convention and has been duly authorized, in accordance with its internal procedures, to sign, ratify, accept, approve or accede to this Convention;

(*k*) "developed country Parties" means developed country Parties and regional economic integration organizations constituted by developed countries.

Article 2. Objective—(1) The objective of this Convention is to combat desertification and mitigate the effects of drought in countries experiencing serious drought and/or desertification, particularly in Africa, through effective action at all levels, supported by international cooperation and partnership arrangements, in the framework of an integrated approach which is consistent with Agenda 21, with a view to contributing to the achievement of sustainable development in affected areas.

(2) Achieving this objective will involve long-term integrated strategies that focus simultaneously, in affected areas, on improved productivity of land, and the rehabilitation, conservation and sustainable management of land and water resources, leading to improved living conditions, in particular at the community level.

Article 3. Principles—In order to achieve the objective of this Convention and to implement its provisions, the Parties shall be guided, *inter alia*, by the following:

(*a*) the Parties should ensure that decisions on the design and implementation of programmes to combat desertification and/or mitigate the effects of drought are taken with the participation of populations and local communities and that an enabling environment is created at higher levels to facilitate action at national and local levels;

(*b*) the Parties should, in a spirit of international solidarity and partnership, improve cooperation and coordination at subregional, regional and international levels, and better focus financial, human, organizational and technical resources where they are needed;

(*c*) the Parties should develop, in a spirit of partnership, cooperation among all levels of government, communities, non-governmental organizations and landholders to establish a better understanding of the nature and value of land and scarce water resources in affected areas and to work towards their sustainable use; and

(*d*) the Parties should take into full consideration the special needs and circumstances of affected developing country Parties, particularly the least developed among them.

PART II
GENERAL PROVISIONS

Article 4. General obligations—(1) The Parties shall implement their obligations under this Convention, individually or jointly, either through existing or prospective bilateral and multilateral arrangements or a combination thereof, as appropriate, emphasizing the need to coordinate efforts and develop a coherent long-term strategy at all levels.

(2) In pursuing the objective of this Convention, the Parties shall:

(*a*) adopt an integrated approach addressing the physical, biological and socio-economic aspects of the processes of desertification and drought;

(*b*) give due attention, within the relevant international and regional bodies, to the situation of affected developing country Parties with regard to international trade, marketing arrangements and debt with a view to establishing an enabling international economic environment conducive to the promotion of sustainable development;

(*c*) integrate strategies for poverty eradication into efforts to combat desertification and mitigate the effects of drought;

(*d*) promote cooperation among affected country Parties in the fields of environmental protection and the conservation of land and water resources, as they relate to desertification and drought;

(*e*) strengthen subregional, regional and international cooperation;

(*f*) cooperate within relevant intergovernmental organizations;

(*g*) determine institutional mechanisms, if appropriate, keeping in mind the need to avoid duplication; and

(*h*) promote the use of existing bilateral and multilateral financial mechanisms and arrangements that mobilize and channel substantial financial resources to affected developing country Parties in combating desertification and mitigating the effects of drought.

(3) Affected developing country Parties are eligible for assistance in the implementation of the Convention.

Article 5. Obligations of affected country Parties—In addition to their obligations pursuant to article 4, affected country Parties undertake to:

(*a*) give due priority to combating desertification and mitigating the effects of drought, and allocate adequate resources in accordance with their circumstances and capabilities;

(*b*) establish strategies and priorities, within the framework of sustainable development plans and/or policies, to combat desertification and mitigate the effects of drought;

(*c*) address the underlying causes of desertification and pay special attention to the socio-economic factors contributing to desertification processes;

(*d*) promote awareness and facilitate the participation of local populations, particularly women and youth, with the support of non-governmental organizations, in efforts to combat desertification and mitigate the effects of drought; and (*e*) provide an enabling environment by strengthening, as appropriate, relevant existing legislation and, where they do not exist, enacting new laws and establishing long-term policies and action programmes.

Article 6. Obligations of developed country Parties—In addition to their general obligations pursuant to article 4, developed country Parties undertake to:

(*a*) actively support, as agreed, individually or jointly, the efforts of affected developing country Parties, particularly those in Africa, and the least developed countries, to combat desertification and mitigate the effects of drought;

(*b*) provide substantial financial resources and other forms of support to assist affected developing country Parties, particularly those in Africa, effectively to develop and implement their own long-term plans and strategies to combat desertification and mitigate the effects of drought;

(*c*) promote the mobilization of new and additional funding pursuant to article 20, paragraph 2 (*b*);

(*d*) encourage the mobilization of funding from the private sector and other non-governmental sources; and

(*e*) promote and facilitate access by affected country Parties, particularly affected developing country Parties, to appropriate technology, knowledge and know-how.

Article 7. Priority for Africa—In implementing this Convention, the Parties shall give priority to affected African country Parties, in the light of the particular situation prevailing in that region, while not neglecting affected developing country Parties in other regions.

Article 8. Relationship with other conventions—(1) The Parties shall encourage the coordination of activities carried out under this Convention and, if they are Parties to them, under other relevant international agreements, particularly the United Nations Framework Convention on Climate Change and the Convention on Biological Diversity, in order to derive maximum benefit from activities under each agreement while avoiding duplication of effort. The Parties shall encourage the conduct of joint programmes, particularly in the fields of research, training, systematic observation and information collection and exchange, to the extent that such activities may contribute to achieving the objectives of the agreements concerned.

(2) The provisions of this Convention shall not affect the rights and obligations of any Party deriving from a bilateral, regional or international agreement into which it has entered prior to the entry into force of this Convention for it.

PART III
ACTION PROGRAMMES, SCIENTIFIC AND TECHNICAL COOPERATION AND SUPPORTING MEASURES

Section 1: Action Programmes

Article 9. Basic approach—(1) In carrying out their obligations pursuant to article 5, affected developing country Parties and any other affected country Party in the framework of its regional implementation annex or, otherwise, that has notified the Permanent Secretariat in writing of its intention to prepare a national action programme, shall, as appropriate, prepare, make public and implement national action programmes, utilizing and building, to the extent possible, on existing relevant successful plans and programmes, and subregional and regional action programmes, as the central element of the strategy to combat desertification and mitigate the effects of drought. Such programmes shall be updated through a continuing participatory process on the basis of lessons from field action, as well as the results of research. The preparation of national action programmes shall be closely interlinked with other efforts to formulate national policies for sustainable development.

(2) In the provision by developed country Parties of different forms of assistance under the terms of article 6, priority shall be given to supporting, as agreed, national, subregional and regional action programmes of affected developing country Parties, particularly those in Africa, either directly or through relevant multilateral organizations or both.

(3) The Parties shall encourage organs, funds and programmes of the United Nations system and other relevant intergovernmental organizations, academic institutions, the scientific community and non-governmental organizations in a position to cooperate, in accordance with their mandates and capabilities, to support the elaboration, implementation and follow-up of action programmes.

Article 10. National action programmes—(1) The purpose of national action programmes is to identify the factors contributing to desertification and practical measures necessary to combat desertification and mitigate the effects of drought.

(2) National action programmes shall specify the respective roles of government, local communities and land users and the resources available and needed. They shall, *inter alia*:

(*a*) incorporate long-term strategies to combat desertification and mitigate the effects of drought, emphasize implementation and be integrated with national policies for sustainable development;

(*b*) allow for modifications to be made in response to changing circumstances and be sufficiently flexible at the local level to cope with different socio-economic, biological and geo-physical conditions;

(*c*) give particular attention to the implementation of preventive measures for lands that are not yet degraded or which are only slightly degraded;

(*d*) enhance national climatological, meteorological and hydrological capabilities and the means to provide for drought early warning;

(*e*) promote policies and strengthen institutional frameworks which develop cooperation and coordination, in a spirit of partnership, between the donor community, governments at all levels, local populations and community groups, and facilitate access by local populations to appropriate information and technology;

(*f*) provide for effective participation at the local, national and regional levels of non-governmental organizations and local populations, both women and men, particularly resource users, including farmers and pastoralists and their representative organizations, in policy planning, decision-making, and implementation and review of national action programmes; and

(*g*) require regular review of, and progress reports on, their implementation.

(3) National action programmes may include, *inter alia*, some or all of the following measures to prepare for and mitigate the effects of drought:

(*a*) establishment and/or strengthening, as appropriate, of early warning systems, including local and national facilities and joint systems at the subregional and regional levels, and mechanisms for assisting environmentally displaced persons;

(*b*) strengthening of drought preparedness and management, including drought contingency plans at the local, national, subregional and regional levels, which take into consideration seasonal to interannual climate predictions;

(*c*) establishment and/or strengthening, as appropriate, of food security systems, including storage and marketing facilities, particularly in rural areas;

(*d*) establishment of alternative livelihood projects that could provide incomes in drought prone areas; and

(*e*) development of sustainable irrigation programmes for both crops and livestock.

(4) Taking into account the circumstances and requirements specific to each affected country Party, national action programmes include, as appropriate, *inter alia,* measures in some or all of the following priority fields as they relate to combating desertification and mitigating the effects of drought in affected areas and to their populations: promotion of alternative livelihoods and improvement of national economic environments with a view to strengthening programmes aimed at the eradication of poverty and at ensuring food security; demographic dynamics; sustainable management of natural resources; sustainable agricultural practices; development and efficient use of various energy sources; institutional and legal frameworks; strengthening of capabilities for assessment and systematic observation, including hydrological and meteorological services, and capacity building, education and public awareness.

Article 11. Subregional and regional action programmes— Affected country Parties shall consult and cooperate to prepare, as appropriate, in accordance with relevant regional implementation annexes, subregional and/or regional action programmes to harmonize, complement and increase the efficiency of national programmes. The provisions of article 10 shall apply *mutatis mutandis* to subregional and regional programmes. Such cooperation may include agreed joint programmes for the sustainable management of transboundary natural resources, scientific and technical cooperation, and strengthening of relevant institutions.

Article 12. International cooperation—Affected country Parties, in collaboration with other Parties and the international community, should cooperate to ensure the promotion of an enabling international environment in the implementation of the Convention. Such cooperation should also cover fields of technology transfer as well as scientific research and development, information collection and dissemination and financial resources.

Article 13. Support for the elaboration and implementation of action programmes——(1) Measures to support action programmes pursuant to article 9 include, *inter alia*:

(*a*) financial cooperation to provide predictability for action programmes, allowing for necessary long-term planning;

(*b*) elaboration and use of cooperation mechanisms which better enable support at the local level, including action through non-governmental organizations, in order to promote the replicability of successful pilot programme activities where relevant;

(*c*) increased flexibility in project design, funding and implementation in keeping with the experimental, iterative approach indicated for participatory action at the local community level; and

(*d*) as appropriate, administrative and budgetary procedures that increase the efficiency of cooperation and of support programmes.

(2) In providing such support to affected developing country Parties, priority shall be given to African country Parties and to least developed country Parties.

Article 14. Coordination in the elaboration and implementation of action programmes——(1) The Parties shall work closely together, directly and through relevant intergovernmental organizations, in the elaboration and implementation of action programmes.

(2) The Parties shall develop operational mechanisms, particularly at the national and field levels, to ensure the fullest possible coordination among developed country Parties, developing country Parties and relevant intergovernmental and

non-governmental organizations, in order to avoid duplication, harmonize interventions and approaches, and maximize the impact of assistance. In affected developing country Parties, priority will be given to coordinating activities related to international cooperation in order to maximize the efficient use of resources, to ensure responsive assistance, and to facilitate the implementation of national action programmes and priorities under this Convention.

Article 15. Regional implementation annexes—Elements for incorporation in action programmes shall be selected and adapted to the socio-economic, geographical and climatic factors applicable to affected country Parties or regions, as well as to their level of development. Guidelines for the preparation of action programmes and their exact focus and content for particular subregions and regions are set out in the regional implementation annexes.

Section 2: Scientific and Technical Cooperation

Article 16. Information collection, analysis and exchange—The Parties agree, according to their respective capabilities, to integrate and coordinate the collection, analysis and exchange of relevant short-term and long-term data and information to ensure systematic observation of land degradation in affected areas and to understand better and assess the processes and effects of drought and desertification. This would help accomplish, *inter alia*, early warning and advance planning for periods of adverse climatic variation in a form suited for practical application by users at all levels, including especially local populations. To this end, they shall, as appropriate:

(*a*) facilitate and strengthen the functioning of the global network of institutions and facilities for the collection, analysis and exchange of information, as well as for systematic observation at all levels, which shall, *inter alia*:

(*i*) aim to use compatible standards and systems;

(*ii*) encompass relevant data and stations, including in remote areas;

(*iii*) use and disseminate modern technology for data collection, transmission and assessment on land degradation; and (*iv*) link national, subregional and regional data and information centres more closely with global information sources;

(*b*) ensure that the collection, analysis and exchange of information address the needs of local communities and those of decision makers, with a view to resolving specific problems, and that local communities are involved in these activities;

(*c*) support and further develop bilateral and multilateral programmes and projects aimed at defining, conducting, assessing and financing the collection, analysis and exchange of data and information, including, *inter alia*, integrated sets of physical, biological, social and economic indicators;

(*d*) make full use of the expertise of competent intergovernmental and non-governmental organizations, particularly to disseminate relevant information and experiences among target groups in different regions;

(*e*) give full weight to the collection, analysis and exchange of socio-economic data, and their integration with physical and biological data;

(*f*) exchange and make fully, openly and promptly available information from all publicly available sources relevant to combating desertification and mitigating the effects of drought; and

(*g*) subject to their respective national legislation and/or policies, exchange information on local and traditional knowledge, ensuring adequate protection for it and providing appropriate return from the benefits derived from it, on an equitable basis and on mutually agreed terms, to the local populations concerned.

Article 17. Research and development—(1) The Parties undertake, according to their respective capabilities, to promote technical and scientific cooperation in the fields of combating desertification and mitigating the effects of drought through appropriate national, subregional, regional and international institutions. To this end, they shall support research activities that:

(*a*) contribute to increased knowledge of the processes leading to desertification and drought and the impact of, and distinction between, causal factors, both natural and human, with a view to combating desertification and mitigating the effects of drought, and achieving improved productivity as well as sustainable use and management of resources;

(*b*) respond to well defined objectives, address the specific needs of local populations and lead to the identification and implementation of solutions that improve the living standards of people in affected areas;

(*c*) protect, integrate, enhance and validate traditional and local knowledge, know-how and practices, ensuring, subject to their respective national legislation and/or policies, that the owners of that knowledge will directly benefit on an equitable basis and on mutually agreed terms from any commercial utilization of it or from any technological development derived from that knowledge;

(*d*) develop and strengthen national, subregional and regional research capabilities in affected developing country Parties, particularly in Africa, including the development of local skills and the strengthening of appropriate capacities, especially in countries with a weak research base, giving particular attention to multidisciplinary and participative socio-economic research;

(*e*) take into account, where relevant, the relationship between poverty, migration caused by environmental factors, and desertification;

(*f*) promote the conduct of joint research programmes between national, subregional, regional and international research organizations, in both the public and private sectors, for the development of improved, affordable and accessible technologies for sustainable development through effective participation of local populations and communities; and

(*g*) enhance the availability of water resources in affected areas, by means of, *inter alia*, cloud-seeding.

(2) Research priorities for particular regions and subregions, reflecting different local conditions, should be included in action programmes. The Conference of the Parties shall review research priorities periodically on the advice of the Committee on Science and Technology.

Article 18. Transfer, acquisition, adaptation and development of technology—(1) The Parties undertake, as mutually agreed and in accordance with their respective national legislation and/or policies, to promote, finance and/or facilitate the financing of the transfer, acquisition, adaptation and development of environmentally sound, economically viable and socially acceptable technologies relevant to combating desertification and/or mitigating the effects of drought, with a view to contributing to the achievement of sustainable development in affected areas. Such cooperation shall be conducted bilaterally or multilaterally, as appropriate, making full use of the expertise of intergovernmental and non-governmental organizations. The Parties shall, in particular:

(*a*) fully utilize relevant existing national, subregional, regional and international information systems and clearing-houses for the dissemination of information on available technologies, their sources, their environmental risks and the broad terms under which they may be acquired;

(*b*) facilitate access, in particular by affected developing country Parties, on favourable terms, including on concessional and preferential terms, as mutually agreed, taking into account the need to protect intellectual property rights, to technologies most suitable to practical application for specific needs of local populations, paying special attention to the social, cultural, economic and environmental impact of such technology;

(*c*) facilitate technology cooperation among affected country Parties through financial assistance or other appropriate means;

(*d*) extend technology cooperation with affected developing country Parties, including, where relevant, joint ventures, especially to sectors which foster alternative livelihoods; and

(*e*) take appropriate measures to create domestic market conditions and incentives, fiscal or otherwise, conducive to the development, transfer, acquisition and adaptation of suitable technology, knowledge, know-how and practices, including measures to ensure adequate and effective protection of intellectual property rights.

(2) The Parties shall, according to their respective capabilities, and subject to their respective national legislation and/or policies, protect, promote and use in particular relevant traditional and local technology, knowledge, know-how and practices and, to that end, they undertake to:

(*a*) make inventories of such technology, knowledge, know-how and practices and their potential uses with the participation of local populations, and disseminate such information, where appropriate, in cooperation with relevant intergovernmental and non-governmental organizations;

(*b*) ensure that such technology, knowledge, know-how and practices are adequately protected and that local populations benefit directly, on an equitable basis and as mutually agreed, from any commercial utilization of them or from any technological development derived therefrom;

(*c*) encourage and actively support the improvement and dissemination of such technology, knowledge, know-how and practices or of the development of new technology based on them; and

(*d*) facilitate, as appropriate, the adaptation of such technology, knowledge, know-how and practices to wide use and integrate them with modern technology, as appropriate.

Section 3: Supporting Measures

Article 19. Capacity building, education and public awareness—(1) The Parties recognize the significance of capacity building—that is to say, institution building, training and development of relevant local and national capacities—in efforts to combat desertification and mitigate the effects of drought. They shall promote, as appropriate, capacity-building:

(*a*) through the full participation at all levels of local people, particularly at the local level, especially women and youth, with the cooperation of non-governmental and local organizations;

(*b*) by strengthening training and research capacity at the national level in the field of desertification and drought;

(*c*) by establishing and/or strengthening support and extension services to disseminate relevant technology methods and techniques more effectively, and by training field agents and members of rural organizations in participatory approaches for the conservation and sustainable use of natural resources;

(*d*) by fostering the use and dissemination of the knowledge, know-how and practices of local people in technical cooperation programmes, wherever possible;

(*e*) by adapting, where necessary, relevant environmentally sound technology and traditional methods of agriculture and pastoralism to modern socio-economic conditions;

(*f*) by providing appropriate training and technology in the use of alternative energy sources, particularly renewable energy resources, aimed particularly at reducing dependence on wood for fuel;

(*g*) through cooperation, as mutually agreed, to strengthen the capacity of affected developing country Parties to develop and implement programmes in the field of collection, analysis and exchange of information pursuant to article 16;

(*h*) through innovative ways of promoting alternative livelihoods, including training in new skills;

(*i*) by training of decision makers, managers, and personnel who are responsible for the collection and analysis of data for the dissemination and use of early warning information on drought conditions and for food production;

(*j*) through more effective operation of existing national institutions and legal frameworks and, where necessary, creation of new ones, along with strengthening of strategic planning and management; and

(*k*) by means of exchange visitor programmes to enhance capacity building in affected country Parties through a long-term, interactive process of learning and study.

(2) Affected developing country Parties shall conduct, in cooperation with other Parties and competent intergovernmental and non-governmental organizations, as appropriate, an interdisciplinary review of available capacity and facilities at the local and national levels, and the potential for strengthening them.

(3) The Parties shall cooperate with each other and through competent intergovernmental organizations, as well as with non-governmental organizations, in undertaking and supporting public awareness and educational programmes in both affected and, where relevant, unaffected country Parties to promote understanding of the causes and effects of desertification and drought and of the importance of meeting the objective of this Convention. To that end, they shall:

(*a*) organize awareness campaigns for the general public;

(*b*) promote, on a permanent basis, access by the public to relevant information, and wide public participation in education and awareness activities;

(*c*) encourage the establishment of associations that contribute to public awareness;

(*d*) develop and exchange educational and public awareness material, where possible in local languages, exchange and second experts to train personnel of affected developing country Parties in carrying out relevant education and awareness programmes, and fully utilize relevant educational material available in competent international bodies;

(*e*) assess educational needs in affected areas, elaborate appropriate school curricula and expand, as needed, educational and adult literacy programmes and opportunities for all, in particular for girls and women, on the identification, conservation and sustainable use and management of the natural resources of affected areas; and

(*f*) develop interdisciplinary participatory programmes integrating desertification and drought awareness into educational systems and in non-formal, adult, distance and practical educational programmes.

(4) The Conference of the Parties shall establish and/or strengthen networks of regional education and training centres to combat desertification and mitigate the effects of drought. These networks shall be coordinated by an institution created or designated for that purpose, in order to train scientific, technical and management personnel and to strengthen existing institutions responsible for education and training in affected country Parties, where appropriate, with a view to harmonizing programmes and

to organizing exchanges of experience among them. These networks shall cooperate closely with relevant intergovernmental and non-governmental organizations to avoid duplication of effort.

Article 20. Financial resources—(1) Given the central importance of financing to the achievement of the objective of the Convention, the Parties, taking into account their capabilities, shall make every effort to ensure that adequate financial resources are available for programmes to combat desertification and mitigate the effects of drought.

(2) In this connection, developed country Parties, while giving priority to affected African country Parties without neglecting affected developing country Parties in other regions, in accordance with article 7, undertake to:

(*a*) mobilize substantial financial resources, including grants and concessional loans, in order to support the implementation of programmes to combat desertification and mitigate the effects of drought;

(*b*) promote the mobilization of adequate, timely and predictable financial resources, including new and additional funding from the Global Environment Facility of the agreed incremental costs of those activities concerning desertification that relate to its four focal areas, in conformity with the relevant provisions of the Instrument establishing the Global Environment Facility;

(*c*) facilitate through international cooperation the transfer of technology, knowledge and know-how; and

(*d*) explore, in cooperation with affected developing country Parties, innovative methods and incentives for mobilizing and channelling resources, including those of foundations, non-governmental organizations and other private sector entities, particularly debt swaps and other innovative means which increase financing by reducing the external debt burden of affected developing country Parties, particularly those in Africa.

(3) Affected developing country Parties, taking into account their capabilities, undertake to mobilize adequate financial resources for the implementation of their national action programmes.

(4) In mobilizing financial resources, the Parties shall seek full use and continued qualitative improvement of all national, bilateral

and multilateral funding sources and mechanisms, using consortia, joint programmes and parallel financing, and shall seek to involve private sector funding sources and mechanisms, including those of non-governmental organizations. To this end, the Parties shall fully utilize the operational mechanisms developed pursuant to article 14.

(5) In order to mobilize the financial resources necessary for affected developing country Parties to combat desertification and mitigate the effects of drought, the Parties shall:

(*a*) rationalize and strengthen the management of resources already allocated for combating desertification and mitigating the effects of drought by using them more effectively and efficiently, assessing their successes and shortcomings, removing hindrances to their effective use and, where necessary, reorienting programmes in light of the integrated long-term approach adopted pursuant to this Convention;

(*b*) give due priority and attention within the governing bodies of multilateral financial institutions, facilities and funds, including regional development banks and funds, to supporting affected developing country Parties, particularly those in Africa, in activities which advance implementation of the Convention, notably action programmes they undertake in the framework of regional implementation annexes; and

(*c*) examine ways in which regional and subregional cooperation can be strengthened to support efforts undertaken at the national level.

(6) Other Parties are encouraged to provide, on a voluntary basis, knowledge, know-how and techniques related to desertification and/or financial resources to affected developing country Parties.

(7) The full implementation by affected developing country Parties, particularly those in Africa, of their obligations under the Convention will be greatly assisted by the fulfilment by developed country Parties of their obligations under the Convention, including in particular those regarding financial resources and transfer of technology. In fulfilling their obligations, developed country Parties should take fully into account that economic and social development and poverty

eradication are the first priorities of affected developing country Parties, particularly those in Africa.

Article 21. Financial mechanisms—(1) The Conference of the Parties shall promote the availability of financial mechanisms and shall encourage such mechanisms to seek to maximize the availability of funding for affected developing country Parties, particularly those in Africa, to implement the Convention. To this end, the Conference of the Parties shall consider for adoption *inter alia* approaches and policies that:

(*a*) facilitate the provision of necessary funding at the national, subregional, regional and global levels for activities pursuant to relevant provisions of the Convention;

(*b*) promote multiple-source funding approaches, mechanisms and arrangements and their assessment, consistent with article 20;

(*c*) provide on a regular basis, to interested Parties and relevant intergovernmental and non-governmental organizations, information on available sources of funds and on funding patterns in order to facilitate coordination among them;

(*d*) facilitate the establishment, as appropriate, of mechanisms, such as national desertification funds, including those involving the participation of non-governmental organizations, to channel financial resources rapidly and efficiently to the local level in affected developing country Parties; and

(*e*) strengthen existing funds and financial mechanisms at the subregional and regional levels, particularly in Africa, to support more effectively the implementation of the Convention.

(2) The Conference of the Parties shall also encourage the provision, through various mechanisms within the United Nations system and through multilateral financial institutions, of support at the national, subregional and regional levels to activities that enable developing country Parties to meet their obligations under the Convention.

(3) Affected developing country Parties shall utilize, and where necessary, establish and/or strengthen, national coordinating mechanisms, integrated in national development programmes, that would ensure the efficient use of all available financial resources. They shall also utilize participatory processes

involving non-governmental organizations, local groups and the private sector, in raising funds, in elaborating as well as implementing programmes and in assuring access to funding by groups at the local level. These actions can be enhanced by improved coordination and flexible programming on the part of those providing assistance.

(4) In order to increase the effectiveness and efficiency of existing financial mechanisms, a Global Mechanism to promote actions leading to the mobilization and channelling of substantial financial resources, including for the transfer of technology, on a grant basis, and/or on concessional or other terms, to affected developing country Parties, is hereby established. This Global Mechanism shall function under the authority and guidance of the Conference of the Parties and be accountable to it.

(5) The Conference of the Parties shall identify, at its first ordinary session, an organization to house the Global Mechanism. The Conference of the Parties and the organization it has identified shall agree upon modalities for this Global Mechanism to ensure *inter alia* that such Mechanism:

(*a*) identifies and draws up an inventory of relevant bilateral and multilateral cooperation programmes that are available to implement the Convention;

(*b*) provides advice, on request, to Parties on innovative methods of financing and sources of financial assistance and on improving the coordination of cooperation activities at the national level;

(*c*) provides interested Parties and relevant intergovernmental and non-governmental organizations with information on available sources of funds and on funding patterns in order to facilitate coordination among them; and

(*d*) reports to the Conference of the Parties, beginning at its second ordinary session, on its activities.

(6) The Conference of the Parties shall, at its first session, make appropriate arrangements with the organization it has identified to house the Global Mechanism for the administrative operations of such Mechanism, drawing to the extent possible on existing budgetary and human resources.

(7) The Conference of the Parties shall, at its third ordinary session, review the policies, operational modalities and activities of the Global Mechanism accountable to it pursuant to paragraph

4, taking into account the provisions of article 7. On the basis of this review, it shall consider and take appropriate action.

PART IV
INSTITUTIONS

Article 22. Conference of the Parties—(1) A Conference of the Parties is hereby established.

(2) The Conference of the Parties is the supreme body of the Convention. It shall make, within its mandate, the decisions necessary to promote its effective implementation. In particular, it shall:

(*a*) regularly review the implementation of the Convention and the functioning of its institutional arrangements in the light of the experience gained at the national, subregional, regional and international levels and on the basis of the evolution of scientific and technological knowledge;

(*b*) promote and facilitate the exchange of information on measures adopted by the Parties, and determine the form and timetable for transmitting the information to be submitted pursuant to article 26, review the reports and make recommendations on them;

(*c*) establish such subsidiary bodies as are deemed necessary for the implementation of the Convention;

(*d*) review reports submitted by its subsidiary bodies and provide guidance to them;

(*e*) agree upon and adopt, by consensus, rules of procedure and financial rules for itself and any subsidiary bodies;

(*f*) adopt amendments to the Convention pursuant to articles 30 and 31;

(*g*) approve a programme and budget for its activities, including those of its subsidiary bodies, and undertake necessary arrangements for their financing;

(*h*) as appropriate, seek the cooperation of, and utilize the services of and information provided by, competent bodies or agencies, whether national or international, intergovernmental or non-governmental;

(*i*) promote and strengthen the relationship with other relevant conventions while avoiding duplication of effort; and (*j*) exercise such other functions as may be necessary for the achievement of the objective of the Convention.

(3) The Conference of the Parties shall, at its first session, adopt its own rules of procedure, by consensus, which shall include decision-making procedures for matters not already covered by decision-making procedures stipulated in the Convention. Such procedures may include specified majorities required for the adoption of particular decisions.

(4) The first session of the Conference of the Parties shall be convened by the interim secretariat referred to in article 35 and shall take place not later than one year after the date of entry into force of the Convention. Unless otherwise decided by the Conference of the Parties, the second, third and fourth ordinary sessions shall be held yearly, and thereafter, ordinary sessions shall be held every two years.

(5) Extraordinary sessions of the Conference of the Parties shall be held at such other times as may be decided either by the Conference of the Parties in ordinary session or at the written request of any Party, provided that, within three months of the request being communicated to the Parties by the Permanent Secretariat, it is supported by at least one third of the Parties.

(6) At each ordinary session, the Conference of the Parties shall elect a Bureau. The structure and functions of the Bureau shall be determined in the rules of procedure. In appointing the Bureau, due regard shall be paid to the need to ensure equitable geographical distribution and adequate representation of affected country Parties, particularly those in Africa.

(7) The United Nations, its specialized agencies and any State member thereof or observers thereto not Party to the Convention, may be represented at sessions of the Conference of the Parties as observers. Any body or agency, whether national or international, governmental or non-governmental, which is qualified in matters covered by the Convention, and which has informed the Permanent Secretariat of its wish to be represented at a session of the Conference of the Parties as an observer, may be so admitted unless at least one-third of the Parties present object. The admission and participation of observers shall be subject to the rules of procedure adopted by the Conference of the Parties.

(8) The Conference of the Parties may request competent national and international organizations which have relevant

expertise to provide it with information relevant to article 16, paragraph (*g*), article 17, paragraph 1 (*c*) and article 18, paragraph 2(*b*).

Article 23. Permanent Secretariat—(1) A Permanent Secretariat is hereby established.

(2) The functions of the Permanent Secretariat shall be:

(*a*) to make arrangements for sessions of the Conference of the Parties and its subsidiary bodies established under the Convention and to provide them with services as required;

(*b*) to compile and transmit reports submitted to it;

(*c*) to facilitate assistance to affected developing country Parties, on request, particularly those in Africa, in the compilation and communication of information required under the Convention;

(*d*) to coordinate its activities with the secretariats of other relevant international bodies and conventions;

(*e*) to enter, under the guidance of the Conference of the Parties, into such administrative and contractual arrangements as may be required for the effective discharge of its functions;

(*f*) to prepare reports on the execution of its functions under this Convention and present them to the Conference of the Parties; and

(*g*) to perform such other secretariat functions as may be determined by the Conference of the Parties.

(3) The Conference of the Parties, at its first session, shall designate a Permanent Secretariat and make arrangements for its functioning.

Article 24. Committee on Science and Technology—(1) A Committee on Science and Technology is hereby established as a subsidiary body of the Conference of the Parties to provide it with information and advice on scientific and technological matters relating to combating desertification and mitigating the effects of drought. The Committee shall meet in conjunction with the ordinary sessions of the Conference of the Parties and shall be multidisciplinary and open to the participation of all Parties. It shall be composed of government representatives competent in the relevant fields of expertise. The Conference of the Parties shall decide, at its first session, on the terms of reference of the Committee.

(2) The Conference of the Parties shall establish and maintain a roster of independent experts with expertise and experience in the relevant fields. The roster shall be based on nominations received in writing from the Parties, taking into account the need for a multidisciplinary approach and broad geographical representation.

(3) The Conference of the Parties may, as necessary, appoint ad hoc panels to provide it, through the Committee, with information and advice on specific issues regarding the state of the art in fields of science and technology relevant to combating desertification and mitigating the effects of drought. These panels shall be composed of experts whose names are taken from the roster, taking into account the need for a multidisciplinary approach and broad geographical representation. These experts shall have scientific backgrounds and field experience and shall be appointed by the Conference of the Parties on the recommendation of the Committee. The Conference of the Parties shall decide on the terms of reference and the modalities of work of these panels.

Article 25. Networking of institutions, agencies and bodies—(1) The Committee on Science and Technology shall, under the supervision of the Conference of the Parties, make provision for the undertaking of a survey and evaluation of the relevant existing networks, institutions, agencies and bodies willing to become units of a network. Such a network shall support the implementation of the Convention.

(2) On the basis of the results of the survey and evaluation referred to in paragraph 1, the Committee on Science and Technology shall make recommendations to the Conference of the Parties on ways and means to facilitate and strengthen networking of the units at the local, national and other levels, with a view to ensuring that the thematic needs set out in articles 16 to 19 are addressed.

(3) Taking into account these recommendations, the Conference of the Parties shall:

(*a*) identify those national, subregional, regional and international units that are most appropriate for networking, and recommend operational procedures, and a time frame, for them; and

(*b*) identify the units best suited to facilitating and strengthening such networking at all levels.

PART V
PROCEDURES

Article 26. Communication of information—(1) Each Party shall communicate to the Conference of the Parties for consideration at its ordinary sessions, through the Permanent Secretariat, reports on the measures which it has taken for the implementation of the Convention. The Conference of the Parties shall determine the timetable for submission and the format of such reports.

(2) Affected country Parties shall provide a description of the strategies established pursuant to article 5 and of any relevant information on their implementation.

(3) Affected country Parties which implement action programmes pursuant to articles 9 to 15 shall provide a detailed description of the programmes and of their implementation.

(4) Any group of affected country Parties may make a joint communication on measures taken at the subregional and/or regional levels in the framework of action programmes.

(5) Developed country Parties shall report on measures taken to assist in the preparation and implementation of action programmes, including information on the financial resources they have provided, or are providing, under the Convention.

(6) Information communicated pursuant to paragraphs 1 to 4 shall be transmitted by the Permanent Secretariat as soon as possible to the Conference of the Parties and to any relevant subsidiary body.

(7) The Conference of the Parties shall facilitate the provision to affected developing countries, particularly those in Africa, on request, of technical and financial support in compiling and communicating information in accordance with this article, as well as identifying the technical and financial needs associated with action programmes.

Article 27. Measures to resolve questions on implementation—The Conference of the Parties shall consider and adopt procedures and institutional mechanisms for the resolution of questions that may arise with regard to the implementation of the Convention.

Article 28. Settlement of disputes—(1) Parties shall settle any dispute between them concerning the interpretation or application of the Convention through negotiation or other peaceful means of their own choice.

(2) When ratifying, accepting, approving, or acceding to the Convention, or at any time thereafter, a Party which is not a regional economic integration organization may declare in a written instrument submitted to the Depositary that, in respect of any dispute concerning the interpretation or application of the Convention, it recognizes one or both of the following means of dispute settlement as compulsory in relation to any Party accepting the same obligation:

(*a*) arbitration in accordance with procedures adopted by the Conference of the Parties in an annex as soon as practicable;

(*b*) submission of the dispute to the International Court of Justice.

(3) A Party which is a regional economic integration organization may make a declaration with like effect in relation to arbitration in accordance with the procedure referred to in paragraph 2 (*a*).

(4) A declaration made pursuant to paragraph 2 shall remain in force until it expires in accordance with its terms or until three months after written notice of its revocation has been deposited with the Depositary.

(5) The expiry of a declaration, a notice of revocation or a new declaration shall not in any way affect proceedings pending before an arbitral tribunal or the International Court of Justice unless the Parties to the dispute otherwise agree.

(6) If the Parties to a dispute have not accepted the same or any procedure pursuant to paragraph 2 and if they have not been able to settle their dispute within twelve months following notification by one Party to another that a dispute exists between them, the dispute shall be submitted to conciliation at the request of any Party to the dispute, in accordance with procedures adopted by the Conference of the Parties in an annex as soon as practicable.

Article 29. Status of annexes—(1) Annexes form an integral part of the Convention and, unless expressly provided otherwise, a reference to the Convention also constitutes a reference to its annexes.

(2) The Parties shall interpret the provisions of the annexes in a manner that is in conformity with their rights and obligations under the articles of this Convention.

Article 30. ***Amendments to the Convention***—(1) Any Party may propose amendments to the Convention.

(2) Amendments to the Convention shall be adopted at an ordinary session of the Conference of the Parties. The text of any proposed amendment shall be communicated to the Parties by the Permanent Secretariat at least six months before the meeting at which it is proposed for adoption. The Permanent Secretariat shall also communicate proposed amendments to the signatories to the Convention.

(3) The Parties shall make every effort to reach agreement on any proposed amendment to the Convention by consensus. If all efforts at consensus have been exhausted and no agreement reached, the amendment shall, as a last resort, be adopted by a two-thirds majority vote of the Parties present and voting at the meeting. The adopted amendment shall be communicated by the Permanent Secretariat to the Depositary, who shall circulate it to all Parties for their ratification, acceptance, approval or accession.

(4) Instruments of ratification, acceptance, approval or accession in respect of an amendment shall be deposited with the Depositary. An amendment adopted pursuant to paragraph 3 shall enter into force for those Parties having accepted it on the ninetieth day after the date of receipt by the Depositary of an instrument of ratification, acceptance, approval or accession by at least two thirds of the Parties to the Convention which were Parties at the time of the adoption of the amendment.

(5) The amendment shall enter into force for any other Party on the ninetieth day after the date on which that Party deposits with the Depositary its instrument of ratification, acceptance or approval of, or accession to the said amendment.

(6) For the purposes of this article and article 31, "Parties present and voting" means Parties present and casting an affirmative or negative vote.

Article 31. Adoption and amendment of annexes—(1) Any additional annex to the Convention and any amendment to an annex shall be proposed and adopted in accordance with the procedure for amendment of the Convention set forth in article 30, provided that, in adopting an additional regional implementation annex or amendment to any regional implementation annex, the majority provided for in that article shall include a two-thirds majority vote of the Parties of the region concerned present and voting. The adoption or amendment of an annex shall be communicated by the Depositary to all Parties.

(2) An annex, other than an additional regional implementation annex, or an amendment to an annex, other than an amendment to any regional implementation annex, that has been adopted in accordance with paragraph 1, shall enter into force for all Parties to the Convention six months after the date of communication by the Depositary to such Parties of the adoption of such annex or amendment, except for those Parties that have notified the Depositary in writing within that period of their non-acceptance of such annex or amendment. Such annex or amendment shall enter into force for Parties which withdraw their notification of non-acceptance on the ninetieth day after the date on which withdrawal of such notification has been received by the Depositary.

(3) An additional regional implementation annex or amendment to any regional implementation annex that has been adopted in accordance with paragraph 1, shall enter into force for all Parties to the Convention six months after the date of the communication by the Depositary to such Parties of the adoption of such annex or amendment, except with respect to:

(*a*) any Party that has notified the Depositary in writing, within such six month period, of its non-acceptance of that additional regional implementation annex or of the amendment to the regional implementation annex, in which case such annex or amendment shall enter into force for Parties which withdraw their notification of non-acceptance on the ninetieth day after the date on which withdrawal of such notification has been received by the Depositary; and

(*b*) any Party that has made a declaration with respect to additional regional implementation annexes or amendments to regional implementation annexes in accordance with article 34, paragraph 4, in which case any such annex or amendment shall enter into force for such a Party on the ninetieth day after the date of deposit with the Depositary of its instrument of ratification, acceptance, approval or accession with respect to such annex or amendment.

(4) If the adoption of an annex or an amendment to an annex involves an amendment to the Convention, that annex or amendment to an annex shall not enter into force until such time as the amendment to the Convention enters into force.

Article 32. Right to vote—(1) Except as provided for in paragraph 2, each Party to the Convention shall have one vote.

(2) Regional economic integration organizations, in matters within their competence, shall exercise their right to vote with a number of votes equal to the number of their member States that are Parties to the Convention. Such an organization shall not exercise its right to vote if any of its member States exercises its right, and vice versa.

PART VI
FINAL PROVISIONS

Article 33. Signature—This Convention shall be opened for signature at Paris, on 14-15 October 1994, by States Members of the United Nations or any of its specialized agencies or that are Parties to the Statute of the International Court of Justice and by regional economic integration organizations. It shall remain open for signature, thereafter, at the United Nations Headquarters in New York until 13 October 1995.

Article 34. Ratification, acceptance, approval and accession—(1) The Convention shall be subject to ratification, acceptance, approval or accession by States and by regional economic integration organizations. It shall be open for accession from the day after the date on which the Convention is closed for signature. Instruments of ratification, acceptance, approval or accession shall be deposited with the Depositary.

(2) Any regional economic integration organization which becomes a Party to the Convention without any of its member States being a Party to the Convention shall be bound by all the obligations under the Convention. Where one or more member States of such an organization are also Party to the Convention, the organization and its member States shall decide on their respective responsibilities for the performance of their obligations under the Convention. In such cases, the organization and the member States shall not be entitled to exercise rights under the Convention concurrently.

(3) In their instruments of ratification, acceptance, approval or accession, regional economic integration organizations shall declare the extent of their competence with respect to the matters governed by the Convention. They shall also promptly inform the Depositary,

who shall in turn inform the Parties, of any substantial modification in the extent of their competence.

(4) In its instrument of ratification, acceptance, approval or accession, any Party may declare that, with respect to it, any additional regional implementation annex or any amendment to any regional implementation annex shall enter into force only upon the deposit of its instrument of ratification, acceptance, approval or accession with respect thereto.

Article 35. Interim arrangements—The secretariat functions referred to in article 23 will be carried out on an interim basis by the secretariat established by the General Assembly of the United Nations in its resolution 47/188 of 22 December 1992, until the completion of the first session of the Conference of the Parties.

Article 36. Entry into force—(1) The Convention shall enter into force on the ninetieth day after the date of deposit of the fiftieth instrument of ratification, acceptance, approval or accession.

(2) For each State or regional economic integration organization ratifying, accepting, approving or acceding to the Convention after the deposit of the fiftieth instrument of ratification, acceptance, approval or accession, the Convention shall enter into force on the ninetieth day after the date of deposit by such State or regional economic integration organization of its instrument of ratification, acceptance, approval or accession.

(3) For the purposes of paragraphs 1 and 2, any instrument deposited by a regional economic integration organization shall not be counted as additional to those deposited by States members of the organization.

Article 37. Reservations—No reservations may be made to this Convention.

Article 38. Withdrawal—(1) At any time after three years from the date on which the Convention has entered into force for a Party, that Party may withdraw from the Convention by giving written notification to the Depositary.

(2) Any such withdrawal shall take effect upon expiry of one year from the date of receipt by the Depositary of the notification of withdrawal, or on such later date as may be specified in the notification of withdrawal.

Article 39. Depositary—The Secretary-General of the United Nations shall be the Depositary of the Convention.

Article 40. Authentic texts—The original of the present Convention, of which the Arabic, Chinese, English, French, Russian and Spanish texts are equally authentic, shall be deposited with the Secretary-General of the United Nations.

In witness whereof the undersigned, being duly authorized to that effect, have signed the present Convention.

Done at Paris, this 17th day of June one thousand nine hundred and ninety-four.

Annex I

Regional Implementation Annex for Africa

Article 1. Scope—This Annex applies to Africa, in relation to each Party and in conformity with the Convention, in particular its article 7, for the purpose of combating desertification and/or mitigating the effects of drought in its arid, semi-arid and dry sub-humid areas.

Article 2. Purpose—The purpose of this Annex, at the national, subregional and regional levels in Africa and in the light of its particular conditions, is to:

(*a*) identify measures and arrangements, including the nature and processes of assistance provided by developed country Parties, in accordance with the relevant provisions of the Convention;

(*b*) provide for the efficient and practical implementation of the Convention to address conditions specific to Africa; and

(*c*) promote processes and activities relating to combating desertification and/or mitigating the effects of drought within the arid, semi-arid and dry sub-humid areas of Africa.

Article 3. Particular conditions of the African region—In carrying out their obligations under the Convention, the Parties shall, in the implementation of this Annex, adopt a basic approach that takes into consideration the following particular conditions of Africa:

(*a*) the high proportion of arid, semi-arid and dry sub-humid areas;

(*b*) the substantial number of countries and populations adversely affected by desertification and by the frequent recurrence of severe drought;

(*c*) the large number of affected countries that are landlocked;

(*d*) the widespread poverty prevalent in most affected countries, the large number of least developed countries among them, and their need for significant amounts of external assistance, in the form of grants and loans on concessional terms, to pursue their development objectives;

(*e*) the difficult socio-economic conditions, exacerbated by deteriorating and fluctuating terms of trade, external indebtedness and political instability, which induce internal, regional and international migrations;

(*f*) the heavy reliance of populations on natural resources for subsistence which, compounded by the effects of demographic trends and factors, a weak technological base and unsustainable production practices, contributes to serious resource degradation;

(*g*) the insufficient institutional and legal frameworks, the weak infrastructural base and the insufficient scientific, technical and educational capacity, leading to substantial capacity building requirements; and

(*h*) the central role of actions to combat desertification and/or mitigate the effects of drought in the national development priorities of affected African countries.

Article 4. Commitments and obligations of African country Parties—(1) In accordance with their respective capabilities, African country Parties undertake to:

(*a*) adopt the combating of desertification and/or the mitigation of the effects of drought as a central strategy in their efforts to eradicate poverty;

(*b*) promote regional cooperation and integration, in a spirit of solidarity and partnership based on mutual interest, in programmes and activities to combat desertification and/or mitigate the effects of drought;

(*c*) rationalize and strengthen existing institutions concerned with desertification and drought and involve other existing institutions, as appropriate, in order to make them more effective and to ensure more efficient use of resources;

(*d*) promote the exchange of information on appropriate technology, knowledge, know-how and practices between and among them; and

(*e*) develop contingency plans for mitigating the effects of drought in areas degraded by desertification and/or drought.

(2) Pursuant to the general and specific obligations set out in articles 4 and 5 of the Convention, affected African country Parties shall aim to:

(*a*) make appropriate financial allocations from their national budgets consistent with national conditions and capabilities and reflecting the new priority Africa has accorded to the phenomenon of desertification and/or drought;

(*b*) sustain and strengthen reforms currently in progress toward greater decentralization and resource tenure as well as reinforce participation of local populations and communities; and

(*c*) identify and mobilize new and additional national financial resources, and expand, as a matter of priority, existing national capabilities and facilities to mobilize domestic financial resources.

Article 5. Commitments and obligations of developed country Parties—(1) In fulfilling their obligations pursuant to articles 4, 6 and 7 of the Convention, developed country Parties shall give priority to affected African country Parties and, in this context, shall:

(*a*) assist them to combat desertification and/or mitigate the effects of drought by, *inter alia*, providing and/or facilitating access to financial and/or other resources, and promoting, financing and/or facilitating the financing of the transfer, adaptation and access to appropriate environmental technologies and know-how, as mutually agreed and in accordance with national policies, taking into account their adoption of poverty eradication as a central strategy;

(*b*) continue to allocate significant resources and/or increase resources to combat desertification and/or mitigate the effects of drought; and

(*c*) assist them in strengthening capacities to enable them to improve their institutional frameworks, as well as their scientific and technical capabilities, information collection and analysis, and research and development for the purpose of combating desertification and/or mitigating the effects of drought.

(2) Other country Parties may provide, on a voluntary basis, technology, knowledge and know-how relating to desertification and/or financial resources, to affected African country Parties. The transfer of such knowledge, know-how and techniques is facilitated by international cooperation.

Article 6. Strategic planning framework for sustainable development—(1) National action programmes shall be a central and integral part of a broader process of formulating national policies for the sustainable development of affected African country Parties.

(2) A consultative and participatory process involving appropriate levels of government, local populations, communities and non-governmental organizations shall be undertaken to provide guidance on a strategy with flexible planning to allow maximum participation from local populations and communities. As appropriate, bilateral and multilateral assistance agencies may be involved in this process at the request of an affected African country Party.

Article 7. Timetable for preparation of action programmes—Pending entry into force of this Convention, the African country Parties, in cooperation with other members of the international community, as appropriate, shall, to the extent possible, provisionally apply those provisions of the Convention relating to the preparation of national, subregional and regional action programmes.

Article 8. Content of national action programmes—(1) Consistent with article 10 of the Convention, the overall strategy of national action programmes shall emphasize integrated local development programmes for affected areas, based on participatory mechanisms and on integration of strategies for poverty eradication into efforts to combat desertification and mitigate the effects of drought. The programmes shall aim at strengthening the capacity of local authorities and ensuring the active involvement of local populations, communities and groups, with emphasis on education and training, mobilization of non-governmental organizations with proven expertise and strengthening of decentralized governmental structures.

(2) National action programmes shall, as appropriate, include the following general features:

(*a*) the use, in developing and implementing national action programmes, of past experiences in combating desertification and/or mitigating the effects of drought, taking into account social, economic and ecological conditions;

(*b*) the identification of factors contributing to desertification and/or drought and the resources and capacities available and required, and the setting up of appropriate policies and institutional and other responses and measures necessary to combat those phenomena and/or mitigate their effects; and

(*c*) the increase in participation of local populations and communities, including women, farmers and pastoralists, and delegation to them of more responsibility for management.

(3) National action programmes shall also, as appropriate, include the following:

(*a*) measures to improve the economic environment with a view to eradicating poverty:

(*i*) increasing incomes and employment opportunities, especially for the poorest members of the community, by:

— developing markets for farm and livestock products;

— creating financial instruments suited to local needs;

— encouraging diversification in agriculture and the setting-up of agricultural enterprises; and

— developing economic activities of a para-agricultural or non-agricultural type;

(*ii*) improving the long-term prospects of rural economies by the creation of:

— incentives for productive investment and access to the means of production; and

— price and tax policies and commercial practices that promote growth;

(*iii*) defining and applying population and migration policies to reduce population pressure on land; and (*iv*) promoting the use of drought resistant crops and the application of integrated dry-land farming systems for food security purposes;

(*b*) measures to conserve natural resources:

(*i*) ensuring integrated and sustainable management of natural resources, including:

— agricultural land and pastoral land;
— vegetation cover and wildlife;
— forests;
— water resources; and
— biological diversity;

(*ii*) training with regard to, and strengthening, public awareness and environmental education campaigns and disseminating knowledge of techniques relating to the sustainable management of natural resources; and

(*iii*) ensuring the development and efficient use of diverse energy sources, the promotion of alternative sources of energy, particularly solar energy, wind energy and bio-gas, and specific arrangements for the transfer, acquisition and adaptation of relevant technology to alleviate the pressure on fragile natural resources;

(*c*) measures to improve institutional organization:

(*i*) defining the roles and responsibilities of central government and local authorities within the framework of a land use planning policy;

(*ii*) encouraging a policy of active decentralization, devolving responsibility for management and decision-making to local authorities, and encouraging initiatives and the assumption of responsibility by local communities and the establishment of local structures; and

(*iii*) adjusting, as appropriate, the institutional and regulatory framework of natural resource management to provide security of land tenure for local populations;

(*d*) measures to improve knowledge of desertification:

(*i*) promoting research and the collection, processing and exchange of information on the scientific, technical and socio-economic aspects of desertification;

(*ii*) improving national capabilities in research and in the collection, processing, exchange and analysis of information so as to increase understanding and to translate the results of the analysis into operational terms; and

(*iii*) encouraging the medium and long-term study of:

— socio-economic and cultural trends in affected areas;

— qualitative and quantitative trends in natural resources; and

— the interaction between climate and desertification; and

(*e*) measures to monitor and assess the effects of drought:

(*i*) developing strategies to evaluate the impacts of natural climate variability on regional drought and desertification and/or to utilize predictions of climate variability on seasonal to interannual time scales in efforts to mitigate the effects of drought;

(*ii*) improving early warning and response capacity, efficiently managing emergency relief and food aid, and improving food stocking and distribution systems, cattle protection schemes and public works and alternative livelihoods for drought prone areas; and

(*iii*) monitoring and assessing ecological degradation to provide reliable and timely information on the process and dynamics of resource degradation in order to facilitate better policy formulations and responses.

Article 9. Preparation of national action programmes and implementation and evaluation indicators—Each affected African country Party shall designate an appropriate national coordinating body to function as a catalyst in the preparation, implementation and evaluation of its national action programme. This coordinating body shall, in the light of article 3 and as appropriate:

(*a*) undertake an identification and review of actions, beginning with a locally driven consultation process, involving local populations and communities and with the cooperation of local administrative authorities, developed country Parties and intergovernmental and non-governmental organizations, on the basis of initial consultations of those concerned at the national level;

(*b*) identify and analyze the constraints, needs and gaps affecting development and sustainable land use and recommend practical measures to avoid duplication by making full use of relevant ongoing efforts and promote implementation of results;

(*c*) facilitate, design and formulate project activities based on interactive, flexible approaches in order to ensure active participation of the population in affected areas, to minimize the negative impact of such activities, and to identify and prioritize requirements for financial assistance and technical cooperation;

(*d*) establish pertinent, quantifiable and readily verifiable indicators to ensure the assessment and evaluation of national action programmes, which encompass actions in the short, medium and long-terms, and of the implementation of such programmes; and

(*e*) prepare progress reports on the implementation of the national action programmes.

Article 10. Organizational framework of subregional action programmes—(1) Pursuant to article 4 of the Convention, African country Parties shall cooperate in the preparation and implementation of subregional action programmes for central, eastern, northern, southern and western Africa and, in that regard, may delegate the following responsibilities to relevant subregional intergovernmental organizations:

(*a*) acting as focal points for preparatory activities and coordinating the implementation of the subregional action programmes;

(*b*) assisting in the preparation and implementation of national action programmes;

(*c*) facilitating the exchange of information, experience and know-how as well as providing advice on the review of national legislation; and

(*d*) any other responsibilities relating to the implementation of subregional action programmes.

(2) Specialized subregional institutions may provide support, upon request, and/or be entrusted with the responsibility to coordinate activities in their respective fields of competence.

Article 11. Content and preparation of subregional action programmes—Subregional action programmes shall focus on issues that are better addressed at the subregional level. They shall establish, where necessary, mechanisms for the management of shared natural resources. Such mechanisms shall effectively handle transboundary problems associated with desertification and/or drought and shall provide support for the harmonious implementation of national action programmes. Priority areas for subregional action programmes shall, as appropriate, focus on:

(*a*) joint programmes for the sustainable management of transboundary natural resources through bilateral and multilateral mechanisms, as appropriate;

(*b*) coordination of programmes to develop alternative energy sources;

(*c*) cooperation in the management and control of pests as well as of plant and animal diseases;

(*d*) capacity building, education and public awareness activities that are better carried out or supported at the subregional level;

(*e*) scientific and technical cooperation, particularly in the climatological, meteorological and hydrological fields, including networking for data collection and assessment, information sharing and project monitoring, and coordination and prioritization of research and development activities;

(*f*) early warning systems and joint planning for mitigating the effects of drought, including measures to address the problems resulting from environmentally induced migrations;

(*g*) exploration of ways of sharing experiences, particularly regarding participation of local populations and communities, and creation of an enabling environment for improved land use management and for use of appropriate technologies;

(*h*) strengthening of the capacity of subregional organizations to coordinate and provide technical services, as well as establishment, reorientation and strengthening of subregional centres and institutions; and

(*i*) development of policies in fields, such as trade, which have impact upon affected areas and populations, including policies for the coordination of regional marketing regimes and for common infrastructure.

Article 12. Organizational framework of the regional action programme—(1) Pursuant to article 11 of the Convention, African country Parties shall jointly determine the procedures for preparing and implementing the regional action programme.

(2) The Parties may provide appropriate support to relevant African regional institutions and organizations to enable them to assist African country Parties to fulfil their responsibilities under the Convention.

Article 13. Content of the regional action programme—The regional action programme includes measures relating to combating desertification and/or mitigating the effects of drought in the following priority areas, as appropriate:

(*a*) development of regional cooperation and coordination of sub-regional action programmes for building regional consensus on key policy areas, including through regular consultations of sub-regional organizations;

(*b*) promotion of capacity building in activities which are better implemented at the regional level;

(*c*) the seeking of solutions with the international community to global economic and social issues that have an impact on affected areas taking into account article 4, paragraph 2 (*b*) of the Convention;

(*d*) promotion among the affected country Parties of Africa and its subregions, as well as with other affected regions, of exchange of information and appropriate techniques, technical know-how and relevant experience; promotion of scientific and technological cooperation particularly in the fields of climatology, meteorology, hydrology, water resource development and alternative energy sources; coordination of sub-regional and regional research activities; and identification of regional priorities for research and development;

(*e*) coordination of networks for systematic observation and assessment and information exchange, as well as their integration into world wide networks; and

(*f*) coordination of and reinforcement of sub-regional and regional early warning systems and drought contingency plans.

Article 14. Financial resources—(1) Pursuant to article 20 of the Convention and article 4, paragraph 2, affected African country Parties shall endeavour to provide a macroeconomic framework conducive to the mobilization of financial resources and shall develop policies and establish procedures to channel resources more effectively to local development programmes, including through non-governmental organizations, as appropriate.

(2) Pursuant to article 21, paragraphs 4 and 5 of the Convention, the Parties agree to establish an inventory of sources of funding at the national, subregional, regional and international levels to ensure the rational use of existing resources and to identify gaps in resource allocation, to facilitate implementation of the action programmes. The inventory shall be regularly reviewed and updated.

(3) Consistent with article 7 of the Convention, the developed country Parties shall continue to allocate significant resources and/or increased resources as well as other forms of assistance to affected African country Parties on the basis of partnership agreements and arrangements referred to in article 18, giving, *inter alia*, due attention to matters related to debt, international trade and marketing arrangements in accordance with article 4, paragraph 2 (*b*) of the Convention.

Article 15. Financial Mechanisms—(1) Consistent with article 7 of the Convention underscoring the priority to affected African country Parties and considering the particular situation prevailing in this region, the Parties shall pay special attention to the implementation in Africa of the provisions of article 21, paragraph 1 (*d*) and (*e*) of the Convention, notably by:

(*a*) facilitating the establishment of mechanisms, such as national desertification funds, to channel financial resources to the local level; and

(*b*) strengthening existing funds and financial mechanisms at the subregional and regional levels.

(2) Consistent with articles 20 and 21 of the Convention, the Parties which are also members of the governing bodies of relevant regional and subregional financial institutions, including the African Development Bank and the African Development Fund, shall promote efforts to give due priority and attention to the activities of those institutions that advance the implementation of this Annex.

(3) The Parties shall streamline, to the extent possible, procedures for channelling funds to affected African country Parties.

Article 16. Technical assistance and cooperation—The Parties undertake, in accordance with their respective capabilities, to rationalize technical assistance to, and cooperation with, African country Parties with a view to increasing project and programme effectiveness by, *inter alia*:

(*a*) limiting the costs of support measures and backstopping, especially overhead costs; in any case, such costs shall only represent an appropriately low percentage of the total cost of the project so as to maximize project efficiency;

(*b*) giving preference to the utilization of competent national experts or, where necessary, competent experts from within the subregion and / or region, in project design, preparation and implementation, and to the building of local expertise where it does not exist; and

(*c*) effectively managing and coordinating, as well as efficiently utilizing, technical assistance to be provided.

Article 17. Transfer, acquisition, adaptation and access to environmentally sound technology—In implementing article 18 of the Convention relating to transfer, acquisition, adaptation and development of technology, the Parties undertake to give priority to African country Parties and, as necessary, to develop with them new models of partnership and cooperation with a view to strengthening capacity building in the fields of scientific research and development and information collection and dissemination to enable them to implement their strategies to combat desertification and mitigate the effects of drought.

Article 18. Coordination and partnership agreements—(1) African country Parties shall coordinate the preparation, negotiation and implementation of national, subregional and regional action programmes. They may involve, as appropriate, other Parties and relevant intergovernmental and non-governmental organizations in this process.

(2) The objectives of such coordination shall be to ensure that financial and technical cooperation is consistent with the Convention and to provide the necessary continuity in the use and administration of resources.

(3) African country Parties shall organize consultative processes at the national, subregional and regional levels. These consultative processes may:

(*a*) serve as a forum to negotiate and conclude partnership agreements based on national, subregional and regional action programmes; and

(*b*) specify the contribution of African country Parties and other members of the consultative groups to the programmes and identify priorities and agreements on implementation and evaluation indicators, as well as funding arrangements for implementation.

(4) The Permanent Secretariat may, at the request of African country Parties, pursuant to article 23 of the Convention, facilitate the convocation of such consultative processes by:

(*a*) providing advice on the organization of effective consultative arrangements, drawing on experiences from other such arrangements;

(*b*) providing information to relevant bilateral and multilateral agencies concerning consultative meetings or processes, and encouraging their active involvement; and

(*c*) providing other information that may be relevant in establishing or improving consultative arrangements.

(5) The subregional and regional coordinating bodies shall, *inter alia*:

(*a*) recommend appropriate adjustments to partnership agreements;

(*b*) monitor, assess and report on the implementation of the agreed subregional and regional programmes; and

(*c*) aim to ensure efficient communication and cooperation among African country Parties.

(6) Participation in the consultative groups shall, as appropriate, be open to Governments, interested groups and donors, relevant organs, funds and programmes of the United Nations system, relevant subregional and regional organizations, and representatives of relevant non-governmental organizations. Participants of each consultative group shall determine the modalities of its management and operation.

(7) Pursuant to article 14 of the Convention, developed country Parties are encouraged to develop, on their own initiative, an informal process of consultation and coordination among themselves, at the

national, subregional and regional levels, and, at the request of an affected African country Party or of an appropriate subregional or regional organization, to participate in a national, subregional or regional consultative process that would evaluate and respond to assistance needs in order to facilitate implementation.

Article 19. Follow-up arrangements—Follow-up of this Annex shall be carried out by African country Parties in accordance with the Convention as follows:

(*a*) at the national level, by a mechanism the composition of which should be determined by each affected African country Party and which shall include representatives of local communities and shall function under the supervision of the national coordinating body referred to in article 9;

(*b*) at the subregional level, by a multidisciplinary scientific and technical consultative committee, the composition and modalities of operation of which shall be determined by the African country Parties of the subregion concerned; and

(*c*) at the regional level, by mechanisms defined in accordance with the relevant provisions of the Treaty establishing the African Economic Community, and by an African Scientific and Technical Advisory Committee.

Annex II

Regional Implementation Annex for Asia

Article 1. Purpose—The purpose of this Annex is to provide guidelines and arrangements for the effective implementation of the Convention in the affected country Parties of the Asian region in the light of its particular conditions.

Article 2. Particular conditions of the Asian region—In carrying out their obligations under the Convention, the Parties shall, as appropriate, take into consideration the following particular conditions which apply in varying degrees to the affected country Parties of the region:

(*a*) the high proportion of areas in their territories affected by, or vulnerable to, desertification and drought and the broad diversity of these areas with regard to climate, topography, land use and socio-economic systems;

(*b*) the heavy pressure on natural resources for livelihoods;

(*c*) the existence of production systems, directly related to widespread poverty, leading to land degradation and to pressure on scarce water resources;

(*d*) the significant impact of conditions in the world economy and social problems such as poverty, poor health and nutrition, lack of food security, migration, displaced persons and demographic dynamics;

(*e*) their expanding, but still insufficient, capacity and institutional frameworks to deal with national desertification and drought problems; and

(*f*) their need for international cooperation to pursue sustainable development objectives relating to combating desertification and mitigating the effects of drought.

Article 3. Framework for national action programmes—(1) National action programmes shall be an integral part of broader national policies for sustainable development of the affected country Parties of the region.

(2) The affected country Parties shall, as appropriate, develop national action programmes pursuant to articles 9 to 11 of the Convention, paying special attention to article 10, paragraph 2 (*f*). As appropriate, bilateral and multilateral cooperation agencies may be involved in this process at the request of the affected country Party concerned.

Article 4. National action programmes—(1) In preparing and implementing national action programmes, the affected country Parties of the region, consistent with their respective circumstances and policies, may, *inter alia*, as appropriate:

(*a*) designate appropriate bodies responsible for the preparation, coordination and implementation of their action programmes;

(*b*) involve affected populations, including local communities, in the elaboration, coordination and implementation of their action programmes through a locally driven consultative process, with the cooperation of local authorities and relevant national and non-governmental organizations;

(*c*) survey the state of the environment in affected areas to assess the causes and consequences of desertification and to determine priority areas for action;

(*d*) evaluate, with the participation of affected populations, past and current programmes for combating desertification and mitigating the effects of drought, in order to design a strategy and elaborate activities in their action programmes;

(*e*) prepare technical and financial programmes based on the information derived from the activities in subparagraphs (*a*) to (*d*);

(*f*) develop and utilize procedures and benchmarks for evaluating implementation of their action programmes;

(*g*) promote the integrated management of drainage basins, the conservation of soil resources, and the enhancement and efficient use of water resources;

(*h*) strengthen and/or establish information, evaluation and follow up and early warning systems in regions prone to desertification and drought, taking account of climatological, meteorological, hydrological, biological and other relevant factors; and

(*i*) formulate in a spirit of partnership, where international cooperation, including financial and technical resources, is involved, appropriate arrangements supporting their action programmes.

(2) Consistent with article 10 of the Convention, the overall strategy of national action programmes shall emphasize integrated local development programmes for affected areas, based on participatory mechanisms and on the integration of strategies for poverty eradication into efforts to combat desertification and mitigate the effects of drought. Sectoral measures in the action programmes shall be grouped in priority fields which take account of the broad diversity of affected areas in the region referred to in article 2 (*a*).

Article 5. Subregional and joint action programmes—(1) Pursuant to article 11 of the Convention, affected country Parties in Asia may mutually agree to consult and cooperate with other Parties, as appropriate, to prepare and implement subregional or joint action programmes, as appropriate, in order to complement, and increase effectiveness in the implementation of, national action programmes. In either case, the relevant Parties may jointly agree to entrust subregional, including bilateral or national organizations, or specialized institutions, with responsibilities relating to the preparation, coordination and implementation of programmes.

Such organizations or institutions may also act as focal points for the promotion and coordination of actions pursuant to articles 16 to 18 of the Convention.

(2) In preparing and implementing subregional or joint action programmes, the affected country Parties of the region shall, *inter alia*, as appropriate:

(*a*) identify, in cooperation with national institutions, priorities relating to combating desertification and mitigating the effects of drought which can better be met by such programmes, as well as relevant activities which could be effectively carried out through them;

(*b*) evaluate the operational capacities and activities of relevant regional, subregional and national institutions;

(*c*) assess existing programmes relating to desertification and drought among all or some parties of the region or subregion and their relationship with national action programmes; and

(*d*) formulate in a spirit of partnership, where international cooperation, including financial and technical resources, is involved, appropriate bilateral and/or multilateral arrangements supporting the programmes.

(3) Subregional or joint action programmes may include agreed joint programmes for the sustainable management of transboundary natural resources relating to desertification, priorities for coordination and other activities in the fields of capacity building, scientific and technical cooperation, particularly drought early warning systems and information sharing, and means of strengthening the relevant subregional and other organizations or institutions.

Article 6. Regional activities—Regional activities for the enhancement of subregional or joint action programmes may include, *inter alia*, measures to strengthen institutions and mechanisms for coordination and cooperation at the national, subregional and regional levels, and to promote the implementation of articles 16 to 19 of the Convention. These activities may also include:

(*a*) promoting and strengthening technical cooperation networks;

(*b*) preparing inventories of technologies, knowledge, know-how and practices, as well as traditional and local technologies and know-how, and promoting their dissemination and use;

(*c*) evaluating the requirements for technology transfer and promoting the adaptation and use of such technologies; and

(*d*) encouraging public awareness programmes and promoting capacity building at all levels, strengthening training, research and development and building systems for human resource development.

Article 7. Financial resources and mechanisms—(1) The Parties shall, in view of the importance of combating desertification and mitigating the effects of drought in the Asian region, promote the mobilization of substantial financial resources and the availability of financial mechanisms, pursuant to articles 20 and 21 of the Convention.

(2) In conformity with the Convention and on the basis of the coordinating mechanism provided for in article 8 and in accordance with their national development policies, affected country Parties of the region shall, individually or jointly:

(*a*) adopt measures to rationalize and strengthen mechanisms to supply funds through public and private investment with a view to achieving specific results in action to combat desertification and mitigate the effects of drought;

(*b*) identify international cooperation requirements in support of national efforts, particularly financial, technical and technological; and

(*c*) promote the participation of bilateral and/or multilateral financial cooperation institutions with a view to ensuring implementation of the Convention.

(3) The Parties shall streamline, to the extent possible, procedures for channelling funds to affected country Parties in the region.

Article 8. Cooperation and coordination mechanisms—(1) Affected country Parties, through the appropriate bodies designated pursuant to article 4, paragraph 1 (*a*), and other Parties in the region, may, as appropriate, set up a mechanism for, *inter alia*, the following purposes:

(*a*) exchange of information, experience, knowledge and know-how;

(*b*) cooperation and coordination of actions, including bilateral and multilateral arrangements, at the subregional and regional levels;

(*c*) promotion of scientific, technical, technological and financial cooperation pursuant to articles 5 to 7;
(*d*) identification of external cooperation requirements; and
(*e*) follow-up and evaluation of the implementation of action programmes.

(2) Affected country Parties, through the appropriate bodies designated pursuant to article 4, paragraph 1 (*a*), and other Parties in the region, may also, as appropriate, consult and coordinate as regards the national, subregional and joint action programmes. They may involve, as appropriate, other Parties and relevant intergovernmental and non-governmental organizations in this process. Such coordination shall, *inter alia*, seek to secure agreement on opportunities for international cooperation in accordance with articles 20 and 21 of the Convention, enhance technical cooperation and channel resources so that they are used effectively.

(3) Affected country Parties of the region shall hold periodic coordination meetings, and the Permanent Secretariat may, at their request, pursuant to article 23 of the Convention, facilitate the convocation of such coordination meetings by:
(*a*) providing advice on the organization of effective coordination arrangements, drawing on experience from other such arrangements;
(*b*) providing information to relevant bilateral and multilateral agencies concerning coordination meetings, and encouraging their active involvement; and
(*c*) providing other information that may be relevant in establishing or improving coordination processes.

Annex III

Regional Implementation Annex for Latin America and the Caribbean

Article 1. Purpose—The purpose of this Annex is to provide general guidelines for the implementation of the Convention in the Latin American and Caribbean region, in light of its particular conditions.

Article 2. Particular conditions of the Latin American and Caribbean region—The Parties shall, in accordance with the provisions of the Convention, take into consideration the following particular conditions of the region:

(*a*) the existence of broad expanses which are vulnerable and have been severely affected by desertification and/or drought and in which diverse characteristics may be observed, depending on the area in which they occur; this cumulative and intensifying process has negative social, cultural, economic and environmental effects which are all the more serious in that the region contains one of the largest resources of biological diversity in the world;

(*b*) the frequent use of unsustainable development practices in affected areas as a result of complex interactions among physical, biological, political, social, cultural and economic factors, including international economic factors such as external indebtedness, deteriorating terms of trade and trade practices which affect markets for agricultural, fishery and forestry products; and

(*c*) a sharp drop in the productivity of ecosystems being the main consequence of desertification and drought, taking the form of a decline in agricultural, livestock and forestry yields and a loss of biological diversity; from the social point of view, the results are impoverishment, migration, internal population movements, and the deterioration of the quality of life; the region will therefore have to adopt an integrated approach to problems of desertification and drought by promoting sustainable development models that are in keeping with the environmental, economic and social situation in each country.

Article 3. Action programmes—(1) In conformity with the Convention, in particular its articles 9 to 11, and in accordance with their national development policies, affected country Parties of the region shall, as appropriate, prepare and implement national action programmes to combat desertification and mitigate the effects of drought as an integral part of their national policies for sustainable development. Subregional and regional programmes may be prepared and implemented in accordance with the requirements of the region.

(2) In the preparation of their national action programmes, affected country Parties of the region shall pay particular attention to article 10, paragraph 2 (*f*) of the Convention.

Article 4. Content of national action programmes—In the light of their respective situations, the affected country Parties of the region may take account, *inter alia*, of the following thematic issues in developing their national strategies for action to combat desertification and/or mitigate the effects of drought, pursuant to article 5 of the Convention:

(*a*) increasing capacities, education and public awareness, technical, scientific and technological cooperation and financial resources and mechanisms;

(*b*) eradicating poverty and improving the quality of human life;

(*c*) achieving food security and sustainable development and management of agricultural, livestock-rearing, forestry and multipurpose activities;

(*d*) sustainable management of natural resources, especially the rational management of drainage basins;

(*e*) sustainable management of natural resources in high-altitude areas;

(*f*) rational management and conservation of soil resources and exploitation and efficient use of water resources;

(*g*) formulation and application of emergency plans to mitigate the effects of drought;

(*h*) strengthening and/or establishing information, evaluation and follow-up and early warning systems in areas prone to desertification and drought, taking account of climatological, meteorological, hydrological, biological, soil, economic and social factors;

(*i*) developing, managing and efficiently using diverse sources of energy, including the promotion of alternative sources;

(*j*) conservation and sustainable use of biodiversity in accordance with the provisions of the Convention on Biological Diversity;

(*k*) consideration of demographic aspects related to desertification and drought; and

(*l*) establishing or strengthening institutional and legal frameworks permitting application of the Convention and aimed, *inter alia*, at decentralizing administrative structures and functions relating to desertification and drought, with the participation of affected communities and society in general.

Article 5. Technical, scientific and technological cooperation—In conformity with the Convention, in particular its articles 16 to 18, and on the basis of the coordinating mechanism provided for in article 7, affected country Parties of the region shall, individually or jointly:

(*a*) promote the strengthening of technical cooperation networks and national, subregional and regional information systems, as well as their integration, as appropriate, in worldwide sources of information;

(*b*) prepare an inventory of available technologies and know-how and promote their dissemination and use;

(*c*) promote the use of traditional technology, knowledge, know-how and practices pursuant to article 18, paragraph 2 (*b*), of the Convention;

(*d*) identify transfer of technology requirements; and

(*e*) promote the development, adaptation, adoption and transfer of relevant existing and new environmentally sound technologies.

Article 6. Financial resources and mechanisms—In conformity with the Convention, in particular its articles 20 and 21, on the basis of the coordinating mechanism provided for in article 7 and in accordance with their national development policies, affected country Parties of the region shall, individually or jointly:

(*a*) adopt measures to rationalize and strengthen mechanisms to supply funds through public and private investment with a view to achieving specific results in action to combat desertification and mitigate the effects of drought;

(*b*) identify international cooperation requirements in support of national efforts; and

(*c*) promote the participation of bilateral and/or multilateral financial cooperation institutions with a view to ensuring implementation of the Convention.

Article 7. Institutional framework—(1) In order to give effect to this Annex, affected country Parties of the region shall:

(*a*) establish and/or strengthen national focal points to coordinate action to combat desertification and/or mitigate the effects of drought; and

(*b*) set up a mechanism to coordinate the national focal points for the following purposes:

(*i*) exchanges of information and experience;

(*ii*) coordination of activities at the subregional and regional levels;

(*iii*) promotion of technical, scientific, technological and financial cooperation;

(*iv*) identification of external cooperation requirements; and

(*v*) follow-up and evaluation of the implementation of action programmes.

(2) Affected country Parties of the region shall hold periodic coordination meetings and the Permanent Secretariat may, at their request, pursuant to article 23 of the Convention, facilitate the convocation of such coordination meetings, by:

(*a*) providing advice on the organization of effective coordination arrangements, drawing on experience from other such arrangements;

(*b*) providing information to relevant bilateral and multilateral agencies concerning coordination meetings, and encouraging their active involvement; and

(*c*) providing other information that may be relevant in establishing or improving coordination processes.

Annex IV

Regional Implementation Annex for the Northern Mediterranean

Article 1. Purpose—The purpose of this Annex is to provide guidelines and arrangements necessary for the effective implementation of the Convention in affected country Parties of the northern Mediterranean region in the light of its particular conditions.

Article 2. Particular conditions of the northern Mediterranean region—The particular conditions of the northern Mediterranean region referred to in article 1 include:

(*a*) semi-arid climatic conditions affecting large areas, seasonal droughts, very high rainfall variability and sudden and high-intensity rainfall;

(*b*) poor and highly erodible soils, prone to develop surface crusts;

(*c*) uneven relief with steep slopes and very diversified landscapes;

(*d*) extensive forest coverage losses due to frequent wildfires;

(*e*) crisis conditions in traditional agriculture with associated land abandonment and deterioration of soil and water conservation structures;

(*f*) unsustainable exploitation of water resources leading to serious environmental damage, including chemical pollution, salinization and exhaustion of aquifers; and

(*g*) concentration of economic activity in coastal areas as a result of urban growth, industrial activities, tourism and irrigated agriculture.

Article 3. Strategic planning framework for sustainable development—(1) National action programmes shall be a central and integral part of the strategic planning framework for sustainable development of the affected country Parties of the northern Mediterranean.

(2) A consultative and participatory process, involving appropriate levels of government, local communities and non-governmental organizations, shall be undertaken to provide guidance on a strategy with flexible planning to allow maximum local participation, pursuant to article 10, paragraph 2 (*f*) of the Convention.

Article 4. Obligation to prepare national action programmes and timetable—Affected country Parties of the northern Mediterranean region shall prepare national action programmes and, as appropriate, subregional, regional or joint action programmes. The preparation of such programmes shall be finalized as soon as practicable.

Article 5. Preparation and implementation of national action programmes—In preparing and implementing national action programmes pursuant to articles 9 and 10 of the Convention, each affected country Party of the region shall, as appropriate:

(*a*) designate appropriate bodies responsible for the preparation, coordination and implementation of its programme;

(*b*) involve affected populations, including local communities, in the elaboration, coordination and implementation of the programme through a locally driven consultative process, with the cooperation of local authorities and relevant non-governmental organizations;

(*c*) survey the state of the environment in affected areas to assess the causes and consequences of desertification and to determine priority areas for action;

(*d*) evaluate, with the participation of affected populations, past and current programmes in order to design a strategy and elaborate activities in the action programme;

(*e*) prepare technical and financial programmes based on the information gained through the activities in subparagraphs (*a*) to (*d*); and

(*f*) develop and utilize procedures and benchmarks for monitoring and evaluating the implementation of the programme.

Article 6. Content of national action programmes—Affected country Parties of the region may include, in their national action programmes, measures relating to:

(*a*) legislative, institutional and administrative areas;

(*b*) land use patterns, management of water resources, soil conservation, forestry, agricultural activities and pasture and range management;

(*c*) management and conservation of wildlife and other forms of biological diversity;

(*d*) protection against forest fires;

(*e*) promotion of alternative livelihoods; and

(*f*) research, training and public awareness.

Article 7. Subregional, regional and joint action programmes—(1) Affected country Parties of the region may, in accordance with article 11 of the Convention, prepare and implement subregional and/or regional action programmes in order to complement and increase the efficiency of national action programmes. Two or more affected country Parties of the region, may similarly agree to prepare a joint action programme between or among them.

(2) The provisions of articles 5 and 6 shall apply *mutatis mutandis* to the preparation and implementation of subregional, regional and joint action programmes. In addition, such programmes may include the conduct of research and development activities concerning selected ecosystems in affected areas.

(3) In preparing and implementing subregional, regional or joint action programmes, affected country Parties of the region shall, as appropriate:

(*a*) identify, in cooperation with national institutions, national objectives relating to desertification which can better be met by such programmes and relevant activities which could be effectively carried out through them;

(*b*) evaluate the operational capacities and activities of relevant regional, subregional and national institutions; and

(*c*) assess existing programmes relating to desertification among Parties of the region and their relationship with national action programmes.

Article 8. Coordination of subregional, regional and joint action programmes—Affected country Parties preparing a subregional, regional or joint action programme may establish a coordination committee composed of representatives of each affected country Party concerned to review progress in combating desertification, harmonize national action programmes, make recommendations at the various stages of preparation and implementation of the subregional, regional or joint action programme, and act as a focal point for the promotion and coordination of technical cooperation pursuant to articles 16 to 19 of the Convention.

Article 9. Non-eligibility for financial assistance—In implementing national, subregional, regional and joint action programmes, affected developed country Parties of the region are not eligible to receive financial assistance under this Convention.

Article 10. Coordination with other subregions and regions—Subregional, regional and joint action programmes in the northern Mediterranean region may be prepared and implemented in collaboration with those of other subregions or regions, particularly with those of the subregion of northern Africa.

Annex V

Regional Implementation Annex for Central and Eastern Europe

Article 1. Purpose—The purpose of this Annex is to provide guidelines and arrangements for the effective implementation of the Convention in affected country Parties of the Central and Eastern European region, in the light of its particular conditions.

Article 2. Particular conditions of the central and eastern European region—The particular conditions of the Central and Eastern European region referred to in article 1, which apply in varying degrees to the affected country Parties of the region, include:

(*a*) specific problems and challenges related to the current process of economic transition, including macroeconomic and financial problems and the need for strengthening the social and political framework for economic and market reforms;

(*b*) the variety of forms of land degradation in the different ecosystems of the region, including the effects of drought and the risks of desertification in regions prone to soil erosion caused by water and wind;

(*c*) crisis conditions in agriculture due, *inter alia*, to depletion of arable land, problems related to inappropriate irrigation systems and gradual deterioration of soil and water conservation structures;

(*d*) unsustainable exploitation of water resources leading to serious environmental damage, including chemical pollution, salinisation and exhaustion of aquifers;

(*e*) forest coverage losses due to climatic factors, consequences of air pollution and frequent wildfires;

(*f*) the use of unsustainable development practices in affected areas as a result of complex interactions among physical, biological, political, social and economic factors;

(*g*) the risks of growing economic hardships and deteriorating social conditions in areas affected by land degradation, desertification and drought;

(*h*) the need to review research objectives and the policy and legislative framework for the sustainable management of natural resources; and

(*i*) the opening up of the region to wider international cooperation and the pursuit of broad objectives of sustainable development.

Article 3. Action programmes—(1) National action programmes shall be an integral part of the policy framework for sustainable development and address in an appropriate manner the various forms of land degradation, desertification and drought affecting the Parties of the region.

(2) A consultative and participatory process, involving appropriate levels of government, local communities and non-governmental organizations, shall be undertaken to provide guidance on a strategy with flexible planning to allow maximum local participation, pursuant to article 10, paragraph 2(*f*), of the Convention. As appropriate, bilateral and multilateral cooperation agencies may be involved in this process at the request of the affected country Party concerned.

Article 4. Preparation and implementation of national action programmes—In preparing and implementing national action programmes pursuant to articles 9 and 10 of the Convention, each affected country Party of the region shall, as appropriate:

(*a*) designate appropriate bodies responsible for the preparation, coordination and implementation of its programme;

(*b*) involve affected populations, including local communities, in the elaboration, coordination and implementation of the programme through a locally driven consultative process, with the cooperation of local authorities and relevant non-governmental organizations;

(*c*) survey the state of the environment in affected areas to assess the causes and consequences of desertification and to determine priority areas for action;

(*d*) evaluate, with the participation of affected populations, past and current programmes in order to design a strategy and elaborate actions in the action programme;

(*e*) prepare technical and financial programmes based on the information gained through the activities in subparagraphs (*a*) to (*d*); and

(*f*) develop and utilize procedures and benchmarks for monitoring and evaluating the implementation of the programme.

Article 5. Subregional, regional and joint action programmes—(1) Affected country Parties of the region, in accordance with articles 11 and 12 of the Convention, may prepare and implement subregional and/or regional action programmes in order to complement and increase the effectiveness and efficiency of national action programmes. Two or more affected country

Parties of the region may similarly agree to prepare a joint action programme between or among them.

(2) Such programmes may be prepared and implemented in collaboration with other Parties or regions. The objective of such collaboration would be to secure an enabling international environment and to facilitate financial and/or technical support or other forms of assistance to address more effectively desertification and drought issues at different levels.

(3) The provisions of articles 3 and 4 shall apply, mutatis mutandis, to the preparation and implementation of subregional, regional and joint action programmes. In addition, such programmes may include the conduct of research and development activities concerning selected ecosystems in affected areas.

(4) In preparing and implementing subregional, regional or joint action programmes, affected country Parties of the region shall, as appropriate:

(*a*) identify, in cooperation with national institutions, national objectives relating to desertification which can better be met by such programmes, and relevant activities, which could be effectively carried out through them;

(*b*) evaluate the operational capacities and activities of relevant regional, subregional and national institutions;

(*c*) assess existing programmes relating to desertification among Parties of the region and their relationship with national action programmes; and

(*d*) consider action for the coordination of subregional, regional and joint action programmes, including, as appropriate, the establishment of coordination committees composed of representatives of each affected country Party concerned to review progress in combating desertification, harmonize national action programmes, make recommendations at the various stages of preparation and implementation of the subregional, regional or joint action programmes, and act as focal points for the promotion and coordination of technical cooperation pursuant to articles 16 to 19 of the Convention.

Article 6. Technical, scientific and technological cooperation—In conformity with the objective and principles of the Convention, Parties of the region shall, individually or jointly:

(*a*) promote the strengthening of scientific and technical cooperation networks, of monitoring indicators and of information systems at all levels, as well as their integration, as appropriate, in worldwide systems of information; and

(*b*) promote the development, adaptation and transfer of relevant existing and new environmentally sound technologies within and outside the region.

Article 7. Financial resources and mechanisms—In conformity with the objective and principles of the Convention, affected country Parties of the region shall, individually or jointly:

(*a*) adopt measures to rationalize and strengthen mechanisms to supply funds through public and private investment with a view to achieving concrete results in action to combat land degradation and desertification and mitigate the effects of drought;

(*b*) identify international cooperation requirements in support of national efforts, thereby creating, in particular, an enabling environment for investments and encouraging active investment policies and an integrated approach to effectively combating desertification, including early identification of the problems caused by this process;

(*c*) seek the participation of bilateral and/or multilateral partners and financial cooperation institutions with a view to ensuring implementation of the Convention, including programme activities which take into account the specific needs of affected country Parties of the region; and

(*d*) assess the possible impact of article 2(*a*) on the implementation of articles 6, 13 and 20 and other related provisions of the Convention.

Article 8. Institutional framework—(1) In order to give effect to this Annex, Parties of the region shall:

(*a*) establish and/or strengthen national focal points to coordinate action to combat desertification and/or mitigate the effects of drought; and

(*b*) consider mechanisms to strengthen regional cooperation, as appropriate.

(2) The Permanent Secretariat may, at the request of Parties of the region and pursuant to article 23 of the Convention, facilitate the convocation of coordination meetings in the region by:

(*a*) providing advice on the organization of effective coordination arrangements, drawing on experience from other such arrangements; and

(*b*) providing other information that may be relevant in establishing or improving coordination processes.

Chapter 5

Conservation and Management of Resources for Environment and Development Action Plan as per Agenda 21

PROTECTION OF THE ATMOSPHERE

Introduction

Protection of the atmosphere is a broad and multidimensional endeavour involving various sectors of economic activity. The options and measures described in the present chapter are recommended for consideration and, as appropriate, implementation by Governments and other bodies in their efforts to protect the atmosphere.

It is recognized that many of the issues discussed in this chapter are also addressed in such international agreements as the 1985 Vienna Convention for the Protection of the Ozone Layer, the 1987 Montreal Protocol on Substances that Deplete the Ozone Layer as amended, the 1992 United Nations Framework Convention on Climate Change and other international, including regional, instruments. In the case of activities covered by such agreements, it is understood that the recommendations contained in this chapter do not oblige any Government to take measures which exceed the provisions of these legal instruments. However, within the framework of this chapter, Governments are free to carry out additional measures which are consistent with those legal instruments.

It is also recognized that activities that may be undertaken in pursuit of the objectives of this chapter should be coordinated with social and economic development in an integrated manner with a view to avoiding adverse impacts on the latter, taking into full account the legitimate priority needs of developing countries for the achievement of sustained economic growth and the eradication of poverty.

Includes the following four programme areas:

(*a*) Addressing the uncertainties: improving the scientific basis for decision-making;

(*b*) Promoting sustainable development:

  (*i*) Energy development, efficiency and consumption;

  (*ii*) Transportation;

  (*iii*) Industrial development;

  (*iv*) Terrestrial and marine resource development and land use;

(*c*) Preventing stratospheric ozone depletion;

(*d*) Transboundary atmospheric pollution.

A. Addressing the Uncertainties: Improving the Scientific Basis for Decision-making

Basis for Action

Concern about climate change and climate variability, air pollution and ozone depletion has created new demands for scientific, economic and social information to reduce the remaining uncertainties in these fields. Better understanding and prediction of the various properties of the atmosphere and of the affected ecosystems, as well as health impacts and their interactions with socio-economic factors, are needed.

Objectives

The basic objective of this programme area is to improve the understanding of processes that influence and are influenced by the Earth's atmosphere on a global, regional and local scale, including, *inter alia*, physical, chemical, geological, biological, oceanic, hydrological, economic and social processes; to build capacity and enhance international cooperation; and to improve understanding of the economic and social consequences of atmospheric changes and of mitigation and response measures addressing such changes.

Activities

Governments at the appropriate level, with the cooperation of the relevant United Nations bodies and, as appropriate, intergovernmental and non-governmental organizations, and the private sector, should:

(*a*) Promote research related to the natural processes affecting and being affected by the atmosphere, as well as the critical linkages between sustainable development and atmospheric changes, including impacts on human health, ecosystems, economic sectors and society;

(*b*) Ensure a more balanced geographical coverage of the Global Climate Observing System and its components, including the Global Atmosphere Watch, by facilitating, *inter alia*, the establishment and operation of additional systematic observation stations, and by contributing to the development, utilization and accessibility of these databases;

(*c*) Promote cooperation in:

(*i*) The development of early detection systems concerning changes and fluctuations in the atmosphere;

(*ii*) The establishment and improvement of capabilities to predict such changes and fluctuations and to assess the resulting environmental and socio-economic impacts;

(*d*) Cooperate in research to develop methodologies and identify threshold levels of atmospheric pollutants, as well as atmospheric levels of greenhouse gas concentrations, that would cause dangerous anthropogenic interference with the climate system and the environment as a whole, and the associated rates of change that would not allow ecosystems to adapt naturally;

(*e*) Promote, and cooperate in the building of scientific capacities, the exchange of scientific data and information, and the facilitation of the participation and training of experts and technical staff, particularly of developing countries, in the fields of research, data assembly, collection and assessment, and systematic observation related to the atmosphere.

B. Promoting Sustainable Development

1. Energy Development, Efficiency and Consumption

Basis for Action

Energy is essential to economic and social development and improved quality of life. Much of the world's energy, however, is currently produced and consumed in ways that could not be sustained if technology were to remain constant and if overall quantities were to increase substantially. The need to control atmospheric emissions of greenhouse and other gases and substances will increasingly need to be based on efficiency in energy production, transmission, distribution and consumption, and on growing reliance on environmentally sound energy systems, particularly new and renewable sources of energy. All energy sources will need to be used in ways that respect the atmosphere, human health and the environment as a whole.

The existing constraints to increasing the environmentally sound energy supplies required for pursuing the path towards sustainable development, particularly in developing countries, need to be removed.

Objectives

The basic and ultimate objective of this programme area is to reduce adverse effects on the atmosphere from the energy sector by promoting policies or programmes, as appropriate, to increase the contribution of environmentally sound and cost-effective energy systems, particularly new and renewable ones, through less polluting and more efficient energy production, transmission, distribution and use. This objective should reflect the need for equity, adequate energy supplies and increasing energy consumption in developing countries, and should take into consideration the situations of countries that are highly dependent on income generated from the production, processing and export, and/or consumption of fossil fuels and associated energy-intensive products and/or the use of fossil fuels for which countries have serious difficulties in switching to alternatives, and the situations of countries highly vulnerable to adverse effects of climate change.

Activities

Governments at the appropriate level, with the cooperation of the relevant United Nations bodies and, as appropriate, intergovernmental and non-governmental organizations, and the private sector, should:

(*a*) Cooperate in identifying and developing economically viable, environmentally sound energy sources to promote the availability of increased energy supplies to support sustainable development efforts, in particular in developing countries;

(*b*) Promote the development at the national level of appropriate methodologies for making integrated energy, environment and economic policy decisions for sustainable development, *inter alia*, through environmental impact assessments;

(*c*) Promote the research, development, transfer and use of improved energy-efficient technologies and practices, including endogenous technologies in all relevant sectors, giving special attention to the rehabilitation and modernization of power systems, with particular attention to developing countries;

(*d*) Promote the research, development, transfer and use of technologies and practices for environmentally sound energy systems, including new and renewable energy systems, with particular attention to developing countries;

(*e*) Promote the development of institutional, scientific, planning and management capacities, particularly in developing countries, to develop, produce and use increasingly efficient and less polluting forms of energy;

(*f*) Review current energy supply mixes to determine how the contribution of environmentally sound energy systems as a whole, particularly new and renewable energy systems, could be increased in an economically efficient manner, taking into account respective countries' unique social, physical, economic and political characteristics, and examining and implementing, where appropriate, measures to overcome any barriers to their development and use;

(*g*) Coordinate energy plans regionally and subregionally, where applicable, and study the feasibility of efficient distribution of environmentally sound energy from new and renewable energy sources;

(*h*) In accordance with national socio-economic development and environment priorities, evaluate and, as appropriate, promote cost-effective policies or programmes, including administrative, social and economic measures, in order to improve energy efficiency;

(*i*) Build capacity for energy planning and programme management in energy efficiency, as well as for the development, introduction, and promotion of new and renewable sources of energy;

(*j*) Promote appropriate energy efficiency and emission standards or recommendations at the national level, aimed at the development and use of technologies that minimize adverse impacts on the environment;

(*k*) Encourage education and awareness-raising programmes at the local, national, subregional and regional levels concerning energy efficiency and environmentally sound energy systems;

(*l*) Establish or enhance, as appropriate, in cooperation with the private sector, labelling programmes for products to provide decision makers and consumers with information on opportunities for energy efficiency.

2. Transportation

Basis for Action

The transport sector has an essential and positive role to play in economic and social development, and transportation needs will undoubtedly increase. However, since the transport sector is also a source of atmospheric emissions, there is need for a review of existing transport systems and for more effective design and management of traffic and transport systems.

Objectives

The basic objective of this programme area is to develop and promote cost-effective policies or programmes, as appropriate, to limit, reduce or control, as appropriate, harmful emissions into the atmosphere and other adverse environmental effects of the transport sector, taking into account development priorities as well as the specific local and national circumstances and safety aspects.

Activities

Governments at the appropriate level, with the cooperation of the relevant United Nations bodies and, as appropriate, intergovernmental and non-governmental organizations, and the private sector, should:

(*a*) Develop and promote, as appropriate, cost-effective, more efficient, less polluting and safer transport systems, particularly integrated rural and urban mass transit, as well as environmentally sound road networks, taking into account the needs for sustainable social, economic and development priorities, particularly in developing countries;

(*b*) Facilitate at the international, regional, subregional and national levels access to and the transfer of safe, efficient, including resource-efficient, and less polluting transport technologies, particularly to the developing countries, including the implementation of appropriate training programmes;

(*c*) Strengthen, as appropriate, their efforts at collecting, analysing and exchanging relevant information on the relation between environment and transport, with particular emphasis on the systematic observation of emissions and the development of a transport database;

(*d*) In accordance with national socio-economic development and environment priorities, evaluate and, as appropriate, promote cost-effective policies or programmes, including administrative, social and economic measures, in order to encourage use of transportation modes that minimize adverse impacts on the atmosphere;

(*e*) Develop or enhance, as appropriate, mechanisms to integrate transport planning strategies and urban and regional settlement planning strategies, with a view to reducing the environmental impacts of transport;

(*f*) Study, within the framework of the United Nations and its regional commissions, the feasibility of convening regional conferences on transport and the environment.

3. Industrial Development

Basis for Action

Industry is essential for the production of goods and services and is a major source of employment and income, and industrial development as such is essential for economic growth. At the same time, industry is a major resource and materials user and consequently industrial activities result in emissions into the atmosphere and the environment as a whole. Protection of the atmosphere can be enhanced, *inter alia,* by increasing resource and materials efficiency in industry, installing or improving pollution abatement technologies and replacing chlorofluorocarbons (CFCs) and other ozone-depleting substances with appropriate substitutes, as well as by reducing wastes and byproducts.

Objectives

The basic objective of this programme area is to encourage industrial development in ways that minimize adverse impacts on the atmosphere by, *inter alia,* increasing efficiency in the production and consumption by industry of all resources and materials, by improving pollution-abatement technologies and by developing new environmentally sound technologies.

Activities

Governments at the appropriate level, with the cooperation of the relevant United Nations bodies and, as appropriate, intergovernmental and non-governmental organizations, and the private sector, should:

(*a*) In accordance with national socio-economic development and environment priorities, evaluate and, as appropriate, promote cost-effective policies or programmes, including administrative, social and economic measures, in order to minimize industrial pollution and adverse impacts on the atmosphere;

(*b*) Encourage industry to increase and strengthen its capacity to develop technologies, products and processes that are safe, less polluting and make more efficient use of all resources and materials, including energy;

(*c*) Cooperate in the development and transfer of such industrial technologies and in the development of capacities to manage and use such technologies, particularly with respect to developing countries;

(*d*) Develop, improve and apply environmental impact assessments to foster sustainable industrial development;

(*e*) Promote efficient use of materials and resources, taking into account the life cycles of products, in order to realize the economic and environmental benefits of using resources more efficiently and producing fewer wastes;

(*f*) Support the promotion of less polluting and more efficient technologies and processes in industries, taking into account area-specific accessible potentials for energy, particularly safe and renewable sources of energy, with a view to limiting industrial pollution, and adverse impacts on the atmosphere.

4. Terrestrial and Marine Resource Development and Land Use

Basis for Action

Land-use and resource policies will both affect and be affected by changes in the atmosphere. Certain practices related to terrestrial and marine resources and land use can decrease greenhouse gas sinks and increase atmospheric emissions. The loss of biological diversity may reduce the resilience of ecosystems to climatic variations and air pollution damage. Atmospheric changes can have important impacts on forests, biodiversity, and freshwater and marine ecosystems, as well as on economic activities, such as agriculture. Policy objectives in different sectors may often diverge and will need to be handled in an integrated manner.

Objectives

The objectives of this programme area are:

(*a*) To promote terrestrial and marine resource utilization and appropriate land-use practices that contribute to:
   (*i*) The reduction of atmospheric pollution and/or the limitation of anthropogenic emissions of greenhouse gases;
   (*ii*) The conservation, sustainable management and enhancement, where appropriate, of all sinks for greenhouse gases;
   (*iii*) The conservation and sustainable use of natural and environmental resources;

(*b*) To ensure that actual and potential atmospheric changes and their socio-economic and ecological impacts are fully taken into account in planning and implementing policies and programmes concerning terrestrial and marine resources utilization and land-use practices.

Activities

Governments at the appropriate level, with the cooperation of the relevant United Nations bodies and, as appropriate, intergovernmental and non-governmental organizations, and the private sector, should:

(*a*) In accordance with national socio-economic development and environment priorities, evaluate and, as appropriate, promote cost-effective policies or programmes, including administrative, social and economic measures, in order to encourage environmentally sound land-use practices;

(*b*) Implement policies and programmes that will discourage inappropriate and polluting land-use practices and promote sustainable utilization of terrestrial and marine resources;

(*c*) Consider promoting the development and use of terrestrial and marine resources and land-use practices that will be more resilient to atmospheric changes and fluctuations;

(*d*) Promote sustainable management and cooperation in the conservation and enhancement, as appropriate, of sinks and reservoirs of greenhouse gases, including biomass, forests and oceans, as well as other terrestrial, coastal and marine ecosystems.

C. Preventing Stratospheric Ozone Depletion

Basis for Action

Analysis of recent scientific data has confirmed the growing concern about the continuing depletion of the Earth's stratospheric ozone layer by reactive chlorine and bromine from man-made CFCs, halons and related substances. While the 1985 Vienna Convention for the Protection of the Ozone Layer and the 1987 Montreal Protocol on Substances that Deplete the Ozone Layer (as amended in London in 1990) were important steps in international action, the total chlorine loading of the atmosphere of ozone-depleting substances has continued to rise. This can be changed through compliance with the control measures identified within the Protocol.

Objectives

The objectives of this programme area are:

(*a*) To realize the objectives defined in the Vienna Convention and the Montreal Protocol and its 1990 amendments, including the consideration in those instruments of the special needs and conditions of the developing countries and the availability to them of alternatives to substances that deplete the ozone layer. Technologies and natural products that reduce demand for these substances should be encouraged;

(*b*) To develop strategies aimed at mitigating the adverse

effects of ultraviolet radiation reaching the Earth's surface as a consequence of depletion and modification of the stratospheric ozone layer.

Activities

Governments at the appropriate level, with the cooperation of the relevant United Nations bodies and, as appropriate, intergovernmental and non-governmental organizations, and the private sector, should:

(*a*) Ratify, accept or approve the Montreal Protocol and its 1990 amendments; pay their contributions towards the Vienna/Montreal trust funds and the interim multilateral ozone fund promptly; and contribute, as appropriate, towards ongoing efforts under the Montreal Protocol and its implementing mechanisms, including making available substitutes for CFCs and other ozone-depleting substances and facilitating the transfer of the corresponding technologies to developing countries in order to enable them to comply with the obligations of the Protocol;

(*b*) Support further expansion of the Global Ozone Observing System by facilitating—through bilateral and multilateral funding—the establishment and operation of additional systematic observation stations, especially in the tropical belt in the southern hemisphere;

(*c*) Participate actively in the continuous assessment of scientific information and the health and environmental effects, as well as of the technological/economic implications of stratospheric ozone depletion; and consider further actions that prove warranted and feasible on the basis of these assessments;

(*d*) Based on the results of research on the effects of the additional ultraviolet radiation reaching the Earth's surface, consider taking appropriate remedial measures in the fields of human health, agriculture and marine environment;

(*e*) Replace CFCs and other ozone-depleting substances, consistent with the Montreal Protocol, recognizing that a replacement's suitability should be evaluated holistically and not simply based on its contribution to solving one atmospheric or environmental problem.

D. Transboundary Atmospheric Pollution

Basis for Action

Transboundary air pollution has adverse health impacts on humans and other detrimental environmental impacts, such as tree and forest loss and the acidification of water bodies. The geographical distribution of atmospheric pollution monitoring networks is uneven, with the developing countries severely underrepresented. The lack of reliable emissions data outside Europe and North America is a major constraint to measuring transboundary air pollution. There is also insufficient information on the environmental and health effects of air pollution in other regions.

The 1979 Convention on Long-range Transboundary Air Pollution, and its protocols, have established a regional regime in Europe and North America, based on a review process and cooperative programmes for systematic observation of air pollution, assessment and information exchange. These programmes need to be continued and enhanced, and their experience needs to be shared with other regions of the world.

Objectives

The objectives of this programme area are:

(*a*) To develop and apply pollution control and measurement technologies for stationary and mobile sources of air pollution and to develop alternative environmentally sound technologies;

(*b*) To observe and assess systematically the sources and extent of transboundary air pollution resulting from natural processes and anthropogenic activities;

(*c*) To strengthen the capabilities, particularly of developing countries, to measure, model and assess the fate and impacts of transboundary air pollution, through, *inter alia*, exchange of information and training of experts;

(*d*) To develop capabilities to assess and mitigate

transboundary air pollution resulting from industrial and nuclear accidents, natural disasters and the deliberate and/or accidental destruction of natural resources;

(*e*) To encourage the establishment of new and the implementation of existing regional agreements for limiting transboundary air pollution;

(*f*) To develop strategies aiming at the reduction of emissions causing transboundary air pollution and their effects.

Activities

Governments at the appropriate level, with the cooperation of the relevant United Nations bodies and, as appropriate, intergovernmental and non-governmental organizations, the private sector and financial institutions, should:

(*a*) Establish and/or strengthen regional agreements for transboundary air pollution control and cooperate, particularly with developing countries, in the areas of systematic observation and assessment, modelling and the development and exchange of emission control technologies for mobile and stationary sources of air pollution. In this context, greater emphasis should be put on addressing the extent, causes, health and socio-economic impacts of ultraviolet radiation, acidification of the environment and photo-oxidant damage to forests and other vegetation;

(*b*) Establish or strengthen early warning systems and response mechanisms for transboundary air pollution resulting from industrial accidents and natural disasters and the deliberate and/or accidental destruction of natural resources;

(*c*) Facilitate training opportunities and exchange of data, information and national and/or regional experiences;

(*d*) Cooperate on regional, multilateral and bilateral bases to assess transboundary air pollution, and elaborate and implement programmes identifying specific actions to reduce atmospheric emissions and to address their environmental, economic, social and other effects.

Means of Implementation

International and Regional Cooperation

Existing legal instruments have created institutional structures which relate to the purposes of these instruments, and relevant work should primarily continue in those contexts. Governments should continue to cooperate and enhance their cooperation at the regional and global levels, including cooperation within the United Nations system. In this context reference is made to the recommendations in chapter 38 of Agenda 21 (International institutional arrangements).

Capacity-building

Countries, in cooperation with the relevant United Nations bodies, international donors and non-governmental organizations, should mobilize technical and financial resources and facilitate technical cooperation with developing countries to reinforce their technical, managerial, planning and administrative capacities to promote sustainable development and the protection of the atmosphere, in all relevant sectors.

Human Resource Development

Education and awareness-raising programmes concerning the promotion of sustainable development and the protection of the atmosphere need to be introduced and strengthened at the local, national and international levels in all relevant sectors.

Financial and Cost Evaluation

The Conference secretariat has estimated the average total annual cost (1993-2000) of implementing the activities under programme area A to be about $640 million from the international community on grant or concessional terms. These are indicative and order-of-magnitude estimates only and have not been reviewed by Governments. Actual costs and financial terms, including any that are non-concessional, will depend upon, *inter alia*, the specific strategies and programmes Governments decide upon for implementation.

The Conference secretariat has estimated the average total annual cost (1993-2000) of implementing the activities of the four-part programme under programme area B to be about $20 billion from the international community on grant or concessional terms. These are indicative and order-of-magnitude estimates only and have not been reviewed by Governments. Actual costs and financial terms, including any that are non-concessional, will depend upon, *inter alia*, the specific strategies and programmes Governments decide upon for implementation.

The Conference secretariat has estimated the average total annual cost (1993-2000) of implementing the activities under programme area C to be in the range of $160-590 million on grant or concessional terms. These are indicative and order-of-magnitude estimates only and have not been reviewed by Governments. Actual costs and financial terms, including any that are non-concessional, will depend upon, *inter alia*, the specific strategies and programmes Governments decide upon for implementation.

INTEGRATED APPROACH TO THE PLANNING AND MANAGEMENT OF LAND RESOURCES

Land is normally defined as a physical entity in terms of its topography and spatial nature; a broader integrative view also includes natural resources: the soils, minerals, water and biota that the land comprises. These components are organized in ecosystems which provide a variety of services essential to the maintenance of the integrity of life-support systems and the productive capacity of the environment. Land resources are used in ways that take advantage of all these characteristics. Land is a finite resource, while the natural resources it supports can vary over time and according to management conditions and uses. Expanding human requirements and economic activities are placing ever increasing pressures on land resources, creating competition and conflicts and resulting in suboptimal use of both land and land resources. If, in the future, human requirements are to be met in a sustainable manner, it is now essential to resolve these conflicts and move towards more effective and efficient use of land and its natural resources. Integrated physical and land-use planning and management is an eminently practical way to achieve this. By examining all uses of land in an integrated manner, it makes it possible to minimize conflicts, to make the most efficient trade-offs and to link social and economic development with environmental protection and enhancement, thus helping to achieve the objectives of sustainable development. The essence of the integrated approach finds expression in the coordination of the sectoral planning and management activities concerned with the various aspects of land use and land resources.

The present chapter consists of one programme area, the integrated approach to the planning and management of land resources, which deals with the reorganization and, where necessary, some strengthening of the decision-making structure, including existing policies, planning and management procedures and methods that can assist in putting in place an integrated approach to land resources. It does not deal with the operational aspects of planning and management, which are more appropriately dealt with under the relevant sectoral programmes. Since the programme deals with an important cross-sectoral aspect of decision-making for sustainable development, it is closely related to a number of other programmes that deal with that issue directly.

Integrated Approach to the Planning and Management of Land Resources

Basis for Action

Land resources are used for a variety of purposes which interact and may compete with one another; therefore, it is desirable to plan and manage all uses in an integrated manner. Integration should take place at two levels, considering, on the one hand, all environmental, social and economic factors (including, for example, impacts of the various economic and social sectors on the environment and natural resources) and, on the other, all environmental and resource components together (*i.e.*, air, water, biota, land, geological and natural resources). Integrated consideration facilitates appropriate choices and trade-offs, thus maximizing sustainable productivity and use. Opportunities to allocate land to different uses arise in the course of major settlement or development projects or in a sequential fashion as lands become available on the market. This in turn provides opportunities to support traditional patterns of sustainable land management or to assign protected status for conservation of biological diversity or critical ecological services.

A number of techniques, frameworks and processes can be combined to facilitate an integrated approach. They are the indispensable support for the planning and management process, at the national and local level, ecosystem or area levels and for the development of specific plans of action. Many of its elements are already in place but need to be more widely applied, further developed and strengthened. This programme area is concerned primarily with providing a framework that will coordinate decision-making; the content and operational functions are therefore not included here but are dealt with in the relevant sectoral programmes of Agenda 21.

Objectives

The broad objective is to facilitate allocation of land to the uses that provide the greatest sustainable benefits and to promote the transition to a sustainable and integrated management of land resources. In doing so, environmental, social and economic issues should be taken into consideration. Protected areas, private property rights, the rights of indigenous people and their communities and other local communities and the economic role of women in agriculture and rural development, among other issues, should be taken into account. In more specific terms, the objectives are as follows:

(*a*) To review and develop policies to support the best possible use of land and the sustainable management of land resources, by not later than 1996;

(*b*) To improve and strengthen planning, management and evaluation systems for land and land resources, by not later than 2000;

(*c*) To strengthen institutions and coordinating mechanisms for land and land resources, by not later than 1998;

(*d*) To create mechanisms to facilitate the active involvement and participation of all concerned, particularly communities and people at the local level, in decision-making on land use and management, by not later than 1996.

Activities

Management-related Activities

Developing Supportive Policies and Policy Instruments

Governments at the appropriate level, with the support of regional and international organizations, should ensure that policies and policy instruments support the best possible land use and sustainable management of land resources. Particular attention should be given to the role of agricultural land. To do this, they should:

(*a*) Develop integrated goal-setting and policy formulation at the national, regional and local levels that takes into account environmental, social, demographic and economic issues;
(*b*) Develop policies that encourage sustainable land use and management of land resources and take the land resource base, demographic issues and the interests of the local population into account;
(*c*) Review the regulatory framework, including laws, regulations and enforcement procedures, in order to identify improvements needed to support sustainable land use and management of land resources and restricts the transfer of productive arable land to other uses;
(*d*) Apply economic instruments and develop institutional mechanisms and incentives to encourage the best possible land use and sustainable management of land resources;
(*e*) Encourage the principle of delegating policy-making to the lowest level of public authority consistent with effective action and a locally driven approach.

Strengthening Planning and Management Systems

Governments at the appropriate level, with the support of regional and international organizations, should review and, if appropriate, revise planning and management systems to facilitate an integrated approach. To do this, they should:

(*a*) Adopt planning and management systems that facilitate the integration of environmental components such as air, water, land and other natural resources, using landscape ecological planning (LANDEP) or other approaches that focus on, for example, an ecosystem or a watershed;
(*b*) Adopt strategic frameworks that allow the integration of both developmental and environmental goals; examples of these frameworks include sustainable livelihood systems, rural development, the World Conservation Strategy/Caring for the Earth, Primary Environmental Care (PEC) and others;
(*c*) Establish a general framework for land-use and physical planning within which specialized and more detailed sectoral plans (*e.g.*, for protected areas, agriculture, forests, human settlements, rural development) can be developed; establish intersectoral consultative bodies to streamline project planning and implementation;
(*d*) Strengthen management systems for land and natural resources by including appropriate traditional and indigenous methods; examples of these practices include pastoralism, Hema reserves (traditional Islamic land reserves) and terraced agriculture;
(*e*) Examine and, if necessary, establish innovative and flexible approaches to programme funding;
(*f*) Compile detailed land capability inventories to guide sustainable land resources allocation, management and use at the national and local levels.

Promoting Application of Appropriate Tools for Planning and Management

Governments at the appropriate level, with the support of national and international organizations, should promote the improvement, further development and widespread application of planning and management tools that facilitate an integrated and sustainable approach to land and resources. To do this, they should:

(*a*) Adopt improved systems for the interpretation and integrated analysis of data on land use and land resources;

(*b*) Systematically apply techniques and procedures for assessing the environmental, social and economic impacts, risks, costs and benefits of specific actions;

(*c*) Analyse and test methods to include land and ecosystem functions and land resources values in national accounts.

Raising Awareness

Governments at the appropriate level, in collaboration with national institutions and interest groups and with the support of regional and international organizations, should launch awareness-raising campaigns to alert and educate people on the importance of integrated land and land resources management and the role that individuals and social groups can play in it. This should be accompanied by provision of the means to adopt improved practices for land use and sustainable management.

Promoting Public Participation

Governments at the appropriate level, in collaboration with national organizations and with the support of regional and international organizations, should establish innovative procedures, programmes, projects and services that facilitate and encourage the active participation of those affected in the decision-making and implementation process, especially of groups that have, hitherto, often been excluded, such as women, youth, indigenous people and their communities and other local communities.

Data and Information

Strengthening Information Systems

Governments at the appropriate level, in collaboration with national institutions and the private sector and with the support of regional and international organizations, should strengthen the information systems necessary for making decisions and evaluating future changes on land use and management. The needs of both men and women should be taken into account. To do this, they should:

(*a*) Strengthen information, systematic observation and assessment systems for environmental, economic and social data related to land resources at the global, regional, national and local levels and for land capability and land-use and management patterns;

(*b*) Strengthen coordination between existing sectoral data systems on land and land resources and strengthen national capacity to gather and assess data;

(*c*) Provide the appropriate technical information necessary for informed decision-making on land use and management in an accessible form to all sectors of the population, especially to local communities and women;

(*d*) Support low-cost, community-managed systems for the collection of comparable information on the status and processes of change of land resources, including soils, forest cover, wildlife, climate and other elements.

International and Regional Coordination and Cooperation Establishing Regional Machinery

Governments at the appropriate level, with the support of regional and international organizations, should strengthen regional cooperation and exchange of information on land resources. To do this, they should:

(*a*) Study and design regional policies to support programmes for land-use and physical planning;

(*b*) Promote the development of land-use and physical plans in the countries of the region;

(*c*) Design information systems and promote training;

(*d*) Exchange, through networks and other appropriate means, information on experiences with the process and results of integrated and participatory planning and management of land resources at the national and local levels.

Means of Implementation

Financing and Cost Evaluation

The Conference secretariat has estimated the average total annual cost (1993-2000) of implementing the activities of this programme

to be about $50 million from the international community on grant or concessional terms. These are indicative and order-of-magnitude estimates only and have not been reviewed by Governments. Actual costs and financial terms, including any that are non-concessional, will depend upon, *inter alia*, the specific strategies and programmes Governments decide upon for implementation.

Scientific and Technological Means

Enhancing Scientific Understanding of the Land Resources System

Governments at the appropriate level, in collaboration with the national and international scientific community and with the support of appropriate national and international organizations, should promote and support research, tailored to local environments, on the land resources system and the implications for sustainable development and management practices. Priority should be given, as appropriate, to:

(*a*) Assessment of land potential capability and ecosystem functions;
(*b*) Ecosystemic interactions and interactions between land resources and social, economic and environmental systems;
(*c*) Developing indicators of sustainability for land resources, taking into account environmental, economic, social, demographic, cultural and political factors.

Testing Research Findings through Pilot Projects

Governments at the appropriate level, in collaboration with the national and international scientific community and with the support of the relevant international organizations, should research and test, through pilot projects, the applicability of improved approaches to the integrated planning and management of land resources, including technical, social and institutional factors.

Human Resource Development

Enhancing Education and Training

Governments at the appropriate level, in collaboration with the appropriate local authorities, non-governmental organizations and international institutions, should promote the development of the human resources that are required to plan and manage land and land resources sustainably. This should be done by providing incentives for local initiatives and by enhancing local management capacity, particularly of women, through:

(*a*) Emphasizing interdisciplinary and integrative approaches in the curricula of schools and technical, vocational and university training;
(*b*) Training all relevant sectors concerned to deal with land resources in an integrated and sustainable manner;
(*c*) Training communities, relevant extension services, community-based groups and non-governmental organizations on land management techniques and approaches applied successfully elsewhere.

Capacity-building Strengthening Technological Capacity

Governments at the appropriate level, in cooperation with other Governments and with the support of relevant international organizations, should promote focused and concerted efforts for education and training and the transfer of techniques and technologies that support the various aspects of the sustainable planning and management process at the national, state/provincial and local levels. Strengthening institutions.

Governments at the appropriate level, with the support of appropriate international organizations, should:

(*a*) Review and, where appropriate, revise the mandates of institutions that deal with land and natural resources to include explicitly the interdisciplinary integration of environmental, social and economic issues;
(*b*) Strengthen coordinating mechanisms between institutions that deal with land-use and resources management to facilitate integration of sectoral concerns and strategies;
(*c*) Strengthen local decision-making capacity and improve coordination with higher levels.

COMBATING DEFORESTATION

Sustaining the Multiple Roles and Functions of All Types of Forests, Forest Lands and Woodlands

Basis for Action

There are major weaknesses in the policies, methods and mechanisms adopted to support and develop the multiple ecological, economic, social and cultural roles of trees, forests and forest lands. Many developed countries are confronted with the effects of air pollution and fire damage on their forests. More effective measures and approaches are often required at the national level to improve and harmonize policy formulation, planning and programming; legislative measures and instruments; development patterns; participation of the general public, especially women and indigenous people; involvement of youth; roles of the private sector, local organizations, non-governmental organizations and cooperatives; development of technical and multidisciplinary skills and quality of human resources; forestry extension and public education; research capability and support; administrative structures and mechanisms, including intersectoral coordination, decentralization and responsibility and incentive systems; and dissemination of information and public relations. This is especially important to ensure a rational and holistic approach to the sustainable and environmentally sound development of forests. The need for securing the multiple roles of forests and forest lands through adequate and appropriate institutional strengthening has been repeatedly emphasized in many of the reports, decisions and recommendations of FAO, ITTO, UNEP, the World Bank, IUCN and other organizations.

Objectives

The objectives of this programme area are as follows:

(*a*) To strengthen forest-related national institutions, to enhance the scope and effectiveness of activities related to the management, conservation and sustainable development of forests, and to effectively ensure the sustainable utilization and production of forests' goods and services in both the developed and the developing countries; by the year 2000, to strengthen the capacities and capabilities of national institutions to enable them to acquire the necessary knowledge for the protection and conservation of forests, as well as to expand their scope and, correspondingly, enhance the effectiveness of programmes and activities related to the management and development of forests;

(*b*) To strengthen and improve human, technical and professional skills, as well as expertise and capabilities to effectively formulate and implement policies, plans, programmes, research and projects on management, conservation and sustainable development of all types of forests and forest-based resources, and forest lands inclusive, as well as other areas from which forest benefits can be derived.

Activities

Management-related Activities

Governments at the appropriate level, with the support of regional, subregional and international organizations, should, where necessary, enhance institutional capability to promote the multiple roles and functions of all types of forests and vegetation inclusive of other related lands and forest-based resources in supporting sustainable development and environmental conservation in all sectors. This should be done, wherever possible and necessary, by strengthening and/or modifying the existing structures and arrangements, and by improving cooperation and coordination of their respective roles. Some of the major activities in this regard are as follows:

(*a*) Rationalizing and strengthening administrative structures and mechanisms, including provision of adequate levels of staff and allocation of responsibilities, decentralization of decision-making, provision of infrastructural facilities and equipment, intersectoral coordination and an effective system of communication;

(*b*) Promoting participation of the private sector, labour unions, rural cooperatives, local communities, indigenous people, youth, women, user groups and non-governmental organizations in forest-related activities, and access to information and training programmes within the national context;

(*c*) Reviewing and, if necessary, revising measures and programmes relevant to all types of forests and vegetation, inclusive of other related lands and forest-based resources, and relating them to other land uses and development policies and legislation; promoting adequate legislation and other measures as a basis against uncontrolled conversion to other types of land uses;

(*d*) Developing and implementing plans and programmes, including definition of national and, if necessary, regional and subregional goals, programmes and criteria for their implementation and subsequent improvement;

(*e*) Establishing, developing and sustaining an effective system of forest extension and public education to ensure better awareness, appreciation and management of forests with regard to the multiple roles and values of trees, forests and forest lands;

(*f*) Establishing and/or strengthening institutions for forest education and training, as well as forestry industries, for developing an adequate cadre of trained and skilled staff at the professional, technical and vocational levels, with emphasis on youth and women;

(*g*) Establishing and strengthening capabilities for research related to the different aspects of forests and forest products, for example, on the sustainable management of forests, research on biodiversity, on the effects of air-borne pollutants, on traditional uses of forest resources by local populations and indigenous people, and on improving market returns and other nonmarket values from the management of forests.

Data and Information

Governments at the appropriate level, with the assistance and cooperation of international, regional, subregional and bilateral agencies, where relevant, should develop adequate databases and baseline information necessary for planning and programme evaluation. Some of the more specific activities include the following:

(*a*) Collecting, compiling and regularly updating and distributing information on land classification and land use, including data on forest cover, areas suitable for afforestation, endangered species, ecological values, traditional/indigenous land use values, biomass and productivity, correlating demographic, socio-economic and forest resources information at the micro- and macro-levels, and undertaking periodic analyses of forest programmes;

(*b*) Establishing linkages with other data systems and sources relevant to supporting forest management, conservation and development, while further developing or reinforcing existing systems such as geographic information systems, as appropriate;

(*c*) Creating mechanisms to ensure public access to this information.

International and Regional Cooperation and Coordination

Governments at the appropriate level and institutions should cooperate in the provision of expertise and other support and the promotion of international research efforts, in particular with a view to enhancing transfer of technology and specialized training and ensuring access to experiences and research results. There is need for strengthening coordination and improving the performance of existing forest-related international organizations in providing technical cooperation and support to interested countries for the management, conservation and sustainable development of forests.

Means of Implementation

Financial and Cost Evaluation

The secretariat of the Conference has estimated the average total annual cost (1993-2000) of implementing the activities of this

programme to be about $2.5 billion, including about $860 million from the international community on grant or concessional terms. These are indicative and order-ofmagnitude estimates only and have not been reviewed by Governments. Actual costs and financial terms, including any that are non-concessional, will depend upon, *inter alia,* the specific strategies and programmes Governments decide upon for implementation.

Scientific and Technological Means

The planning, research and training activities specified will form the scientific and technological means for implementing the programme, as well as its output. The systems, methodology and knowhow generated by the programme will help improve efficiency. Some of the specific steps involved should include:

(*a*) Analysing achievements, constraints and social issues for supporting programme formulation and implementation;
(*b*) Analysing research problems and research needs, research planning and implementation of specific research projects;
(*c*) Assessing needs for human resources, skill development and training;
(*d*) Developing, testing and applying appropriate methodologies/approaches in implementing forest programmes and plans.

Human Resource Development

The specific components of forest education and training will effectively contribute to human resource development. These include:

(*a*) Launching of graduate and post-graduate degree, specialization and research programmes;
(*b*) Strengthening of pre-service, in-service and extension service training programmes at the technical and vocational levels, including training of trainers/teachers, and developing curriculum and teaching materials/ methods;

(*c*) Special training for staff of national forest-related organizations in aspects such as project formulation, evaluation and periodical evaluations.

Capacity-building

This programme area is specifically concerned with capacity-building in the forest sector and all programme activities specified contribute to that end. In building new and strengthened capacities, full advantage should be taken of the existing systems and experience.

Enhancing the Protection, Sustainable Management and Conservation of All Forests, and the Greening of Degraded Areas, through Forest Rehabilitation, Afforestation, Reforestation and Other Rehabilitative Means

Basis for Action

Forests world wide have been and are being threatened by uncontrolled degradation and conversion to other types of land uses, influenced by increasing human needs; agricultural expansion; and environmentally harmful mismanagement, including, for example, lack of adequate forest-fire control and anti-poaching measures, unsustainable commercial logging, overgrazing and unregulated browsing, harmful effects of airborne pollutants, economic incentives and other measures taken by other sectors of the economy. The impacts of loss and degradation of forests are in the form of soil erosion; loss of biological diversity, damage to wildlife habitats and degradation of watershed areas, deterioration of the quality of life and reduction of the options for development.

The present situation calls for urgent and consistent action for conserving and sustaining forest resources. The greening of suitable areas, in all its component activities, is an effective way of increasing public awareness and participation in protecting and managing forest resources. It should include the consideration of land use and tenure patterns and local needs and should spell out and clarify the specific objectives of the different types of greening activities.

Objectives

The objectives of this programme area are as follows:

(*a*) To maintain existing forests through conservation and management, and sustain and expand areas under forest and tree cover, in appropriate areas of both developed and developing countries, through the conservation of natural forests, protection, forest rehabilitation, regeneration, afforestation, reforestation and tree planting, with a view to maintaining or restoring the ecological balance and expanding the contribution of forests to human needs and welfare;

(*b*) To prepare and implement, as appropriate, national forestry action programmes and/or plans for the management, conservation and sustainable development of forests. These programmes and/or plans should be integrated with other land uses. In this context, country-driven national forestry action programmes and/or plans under the Tropical Forestry Action Programme are currently being implemented in more than 80 countries, with the support of the international community;

(*c*) To ensure sustainable management and, where appropriate, conservation of existing and future forest resources;

(*d*) To maintain and increase the ecological, biological, climatic, socio-cultural and economic contributions of forest resources;

(*e*) To facilitate and support the effective implementation of the non-legally binding authoritative statement of principles for a global consensus on the management, conservation and sustainable development of all types of forests, adopted by the United Nations Conference on Environment and Development, and on the basis of the implementation of these principles to consider the need for and the feasibility of all kinds of appropriate internationally agreed arrangements to promote international cooperation on forest management, conservation and sustainable development of all types of forests, including afforestation, reforestation and rehabilitation.

Activities

Management-related Activities

Governments should recognize the importance of categorizing forests, within the framework of long-term forest conservation and management policies, into different forest types and setting up sustainable units in every region/watershed with a view to securing the conservation of forests. Governments, with the participation of the private sector, non-governmental organizations, local community groups, indigenous people, women, local government units and the public at large, should act to maintain and expand the existing vegetative cover wherever ecologically, socially and economically feasible, through technical cooperation and other forms of support. Major activities to be considered include:

(*a*) Ensuring the sustainable management of all forest ecosystems and woodlands, through improved proper planning, management and timely implementation of silvicultural operations, including inventory and relevant research, as well as rehabilitation of degraded natural forests to restore productivity and environmental contributions, giving particular attention to human needs for economic and ecological services, wood-based energy, agroforestry, non-timber forest products and services, watershed and soil protection, wildlife management, and forest genetic resources;

(*b*) Establishing, expanding and managing, as appropriate to each national context, protected area systems, which includes systems of conservation units for their environmental, social and spiritual functions and values, including conservation of forests in representative ecological systems and landscapes, primary old-growth forests, conservation and management of wildlife, nomination of World Heritage Sites under the World Heritage Convention, as appropriate, conservation of genetic resources, involving *in situ* and *ex situ* measures and undertaking supportive measures to ensure sustainable utilization of biological resources and conservation of biological diversity and the traditional forest habitats of indigenous people, forest dwellers and local communities;

(*c*) Undertaking and promoting buffer and transition zone management;

(*d*) Carrying out revegetation in appropriate mountain areas, highlands, bare lands, degraded farm lands, arid and semi-arid lands and coastal areas for combating desertification and preventing erosion problems and for other protective functions and national programmes for rehabilitation of degraded lands, including community forestry, social forestry, agroforestry and silvipasture, while also taking into account the role of forests as national carbon reservoirs and sinks;

(*e*) Developing industrial and non-industrial planted forests in order to support and promote national ecologically sound afforestation and reforestation/regeneration programmes in suitable sites, including upgrading of existing planted forests of both industrial and nonindustrial and commercial purpose to increase their contribution to human needs and to offset pressure on primary/old growth forests. Measures should be taken to promote and provide intermediate yields and to improve the rate of returns on investments in planted forests, through interplanting and underplanting valuable crops;

(*f*) Developing/strengthening a national and/or master plan for planted forests as a priority, indicating, *inter alia*, the location, scope and species, and specifying areas of existing planted forests requiring rehabilitation, taking into account the economic aspect for future planted forest development, giving emphasis to native species;

(*g*) Increasing the protection of forests from pollutants, fire, pests and diseases and other human made interferences such as forest poaching, mining and unmitigated shifting cultivation, the uncontrolled introduction of exotic plant and animal species, as well as developing and accelerating research for a better understanding of problems relating to the management and regeneration of all types of forests; strengthening and/or establishing appropriate measures to assess and/or check inter-border movement of plants and related materials;

(*h*) Stimulating development of urban forestry for the greening of urban, peri-urban and rural human settlements for amenity, recreation and production purposes and for protecting trees and groves;

(*i*) Launching or improving opportunities for participation of all people, including youth, women, indigenous people and local communities in the formulation, development and implementation of forest-related programmes and other activities, taking due account of the local needs and cultural values;

(*j*) Limiting and aiming to halt destructive shifting cultivation by addressing the underlying social and ecological causes.

Data and Information

Management-related activities should involve collection, compilation and analysis of data/information, including baseline surveys. Some of the specific activities include the following:

(*a*) Carrying out surveys and developing and implementing land-use plans for appropriate greening/planting/afforestation/reforestation/forest rehabilitation;

(*b*) Consolidating and updating land-use and forest inventory and management information for management and land-use planning of wood and non-wood resources, including data on shifting cultivation and other agents of forest destruction;

(*c*) Consolidating information on genetic resources and related biotechnology, including surveys and studies, as necessary;

(*d*) Carrying out surveys and research on local/indigenous knowledge of trees and forests and their uses to improve the planning and implementation of sustainable forest management;

(*e*) Compiling and analysing research data on species/site interaction of species used in planted forests and assessing the potential impact on forests of climatic change, as well as effects of forests on climate, and initiating in-depth studies on the carbon cycle relating to different forest types to provide scientific advice and technical support;

(*f*) Establishing linkages with other data/information sources that relate to sustainable management and use of forests and improving access to data and information;

(*g*) Developing and intensifying research to improve knowledge and understanding of problems and natural mechanisms related to the management and rehabilitation of forests, including research on fauna and its interrelation with forests;

(*h*) Consolidating information on forest conditions and site-influencing immissions and emissions.

International and Regional Cooperation and Coordination

The greening of appropriate areas is a task of global importance and impact. The international and regional community should provide technical cooperation and other means for this programme area. Specific activities of an international nature, in support of national efforts, should include the following:

(*a*) Increasing cooperative actions to reduce pollutants and trans-boundary impacts affecting the health of trees and forests and conservation of representative ecosystems;

(*b*) Coordinating regional and subregional research on carbon sequestration, air pollution and other environmental issues;

(*c*) Documenting and exchanging information/experience for the benefit of countries with similar problems and prospects;

(*d*) Strengthening the coordination and improving the capacity and ability of intergovernmental organizations such as FAO, ITTO, UNEP and UNESCO to provide technical support for the management, conservation and sustainable development of forests, including support for the negotiation of the International Tropical Timber Agreement of 1983, due in 1992/93.

Means of Implementation

Financial and Cost Evaluation

The secretariat of the Conference has estimated the average total annual cost (1993-2000) of implementing the activities of this

programme to be about $10 billion, including about $3.7 billion from the international community on grant or concessional terms. These are indicative and order-of magnitude estimates only and have not been reviewed by Governments. Actual costs and financial terms, including any that are non-concessional, will depend upon, *inter alia*, the specific strategies and programmes Governments decide upon for implementation.

Scientific and Technological Means

Data analysis, planning, research, transfer/development of technology and/or training activities form an integral part of the programme activities, providing the scientific and technological means of implementation. National institutions should:

(*a*) Develop feasibility studies and operational planning related to major forest activities;
(*b*) Develop and apply environmentally sound technology relevant to the various activities listed;
(*c*) Increase action related to genetic improvement and application of biotechnology for improving productivity and tolerance to environmental stress and including, for example, tree breeding, seed technology, seed procurement networks, germ-plasm banks, "in vitro" techniques, and *in situ* and *ex situ* conservation.

Human Resource Development

Essential means for effectively implementing the activities include training and development of appropriate skills, working facilities and conditions, public motivation and awareness. Specific activities include:

(*a*) Providing specialized training in planning, management, environmental conservation, biotechnology etc.;
(*b*) Establishing demonstration areas to serve as models and training facilities;
(*c*) Supporting local organizations, communities, non-governmental organizations and private land owners, in particular women, youth, farmers and indigenous people/

shifting cultivators, through extension and provision of inputs and training.

Capacity-building

National Governments, the private sector, local organizations/ communities, indigenous people, labour unions and non-governmental organizations should develop capacities, duly supported by relevant international organizations, to implement the programme activities. Such capacities should be developed and strengthened in harmony with the programme activities. Capacity-building activities include policy and legal frameworks, national institution building, human resource development, development of research and technology, development of infrastructure, enhancement of public awareness etc.

Promoting Efficient Utilization and Assessment to Recover the Full Valuation of the Goods and Services Provided by Forests, Forest Lands and Woodlands

Basis for Action

The vast potential of forests and forest lands as a major resource for development is not yet fully realized. The improved management of forests can increase the production of goods and services and, in particular, the yield of wood and non-wood forest products, thus helping to generate additional employment and income, additional value through processing and trade of forest products, increased contribution to foreign exchange earnings, and increased return on investment. Forest resources, being renewable, can be sustainably managed in a manner that is compatible with environmental conservation. The implications of the harvesting of forest resources for the other values of the forest should be taken fully into consideration in the development of forest policies. It is also possible to increase the value of forests through non-damaging uses such as eco-tourism and the managed supply of genetic materials. Concerted action is needed in order to increase people's perception of the value of forests and of the benefits they provide. The survival of forests and their continued contribution to human welfare depends to a great extent on succeeding in this endeavour.

Objectives

The objectives of this programme area are as follows:

(*a*) To improve recognition of the social, economic and ecological values of trees, forests and forest lands, including the consequences of the damage caused by the lack of forests; to promote methodologies with a view to incorporating social, economic and ecological values of trees, forests and forest lands into the national economic accounting systems; to ensure their sustainable management in a way that is consistent with land use, environmental considerations and development needs;

(*b*) To promote efficient, rational and sustainable utilization of all types of forests and vegetation inclusive of other related lands and forest-based resources, through the development of efficient forest-based processing industries, value-adding secondary processing and trade in forest products, based on sustainably managed forest resources and in accordance with plans that integrate all wood and non-wood values of forests;

(*c*) To promote more efficient and sustainable use of forests and trees for fuelwood and energy supplies;

(*d*) To promote more comprehensive use and economic contributions of forest areas by incorporating eco-tourism into forest management and planning.

Activities

Management-related Activities

Governments, with the support of the private sector, scientific institutions, indigenous people, non-governmental organizations, cooperatives and entrepreneurs, where appropriate, should undertake the following activities, properly coordinated at the national level, with financial and technical cooperation from international organizations:

(*a*) Carrying out detailed investment studies, supply-demand harmonization and environmental impact analysis to rationalize and improve trees and forest utilization and to develop and establish appropriate incentive schemes and regulatory measures, including tenurial arrangements, to provide a favourable investment climate and promote better management;

(*b*) Formulating scientifically sound criteria and guidelines for the management, conservation and sustainable development of all types of forests;

(*c*) Improving environmentally sound methods and practices of forest harvesting, which are ecologically sound and economically viable, including planning and management, improved use of equipment, storage and transportation to reduce and, if possible, maximize the use of waste and improve value of both wood and non-wood forest products;

(*d*) Promoting the better use and development of natural forests and woodlands, including planted forests, wherever possible, through appropriate and environmentally sound and economically viable activities, including silvicultural practices and management of other plant and animal species;

(*e*) Promoting and supporting the downstream processing of forest products to increase retained value and other benefits;

(*f*) Promoting/popularizing non-wood forest products and other forms of forest resources, apart from fuelwood (*e.g.*, medicinal plants, dyes, fibres, gums, resins, fodder, cultural products, rattan, bamboo) through programmes and social forestry/participatory forest activities, including research on their processing and uses;

(*g*) Developing, expanding and/or improving the effectiveness and efficiency of forest-based processing industries, both wood and non-wood based, involving such aspects as efficient conversion technology and improved sustainable utilization of harvesting and process residues; promoting underutilized species in natural forests through research, demonstration and commercialization; promoting value-adding secondary processing for improved employment, income and retained value; and promoting/improving markets for, and trade in, forest products through relevant institutions, policies and facilities;

(*h*) Promoting and supporting the management of wildlife, as well as eco-tourism, including farming, and encouraging and supporting the husbandry and cultivation of wild species, for improved rural income and employment, ensuring economic and social benefits without harmful ecological impacts;

(*i*) Promoting appropriate small-scale forest-based enterprises for supporting rural development and local entrepreneurship;

(*j*) Improving and promoting methodologies for a comprehensive assessment that will capture the full value of forests, with a view to including that value in the market-based pricing structure of wood and non-wood based products;

(*k*) Harmonizing sustainable development of forests with national development needs and trade policies that are compatible with the ecologically sound use of forest resources, using, for example, the ITTO Guidelines for Sustainable Management of Tropical Forests;

(*l*) Developing, adopting and strengthening national programmes for accounting the economic and non-economic value of forests.

Data and Information

The objectives and management-related activities presuppose data and information analysis, feasibility studies, market surveys and review of technological information. Some of the relevant activities include:

(*a*) Undertaking analysis of supply and demand for forest products and services, to ensure efficiency in their utilization, wherever necessary;

(*b*) Carrying out investment analysis and feasibility studies, including environmental impact assessment, for establishing forest-based processing enterprises;

(*c*) Conducting research on the properties of currently underutilized species for their promotion and commercialization;

(*d*) Supporting market surveys of forest products for trade promotion and intelligence;

(*e*) Facilitating the provision of adequate technological information as a measure to promote better utilization of forest resources.

International and Regional Cooperation and Coordination

Cooperation and assistance of international organizations and the international community in technology transfer, specialization and promotion of fair terms of trade, without resorting to unilateral restrictions and/or bans on forest products contrary to GATT and other multilateral trade agreements, the application of appropriate market mechanisms and incentives will help in addressing global environmental concerns. Strengthening the coordination and performance of existing international organizations, in particular FAO, UNIDO, UNESCO, UNEP, ITC/UNCTAD/GATT, ITTO and ILO, for providing technical assistance and guidance in this programme area is another specific activity.

Means of Implementation

Financial and Cost Evaluation

The secretariat of the Conference has estimated the average total annual cost (1993-2000) of implementing the activities of this programme to be about $18 billion, including about $880 million from the international community on grant or concessional terms. These are indicative and order-ofmagnitude estimates only and have not been reviewed by Governments. Actual costs and financial terms, including any that are non-concessional, will depend upon, *inter alia,* the specific strategies and programmes Governments decide upon for implementation.

Scientific and Technological Means

The programme activities presuppose major research efforts and studies, as well as improvement of technology. This should be coordinated by national Governments, in collaboration with and supported by relevant international organizations and institutions. Some of the specific components include:

(*a*) Research on properties of wood and non-wood products and their uses, to promote improved utilization;
(*b*) Development and application of environmentally sound and less-polluting technology for forest utilization;
(*c*) Models and techniques of outlook analysis and development planning;
(*d*) Scientific investigations on the development and utilization of non-timber forest products;
(*e*) Appropriate methodologies to comprehensively assess the value of forests.

Human Resource Development

The success and effectiveness of the programme area depends on the availability of skilled personnel. Specialized training is an important factor in this regard. New emphasis should be given to the incorporation of women. Human resource development for programme implementation, in quantitative and qualitative terms, should include:

(*a*) Developing required specialized skills to implement the programme, including establishing special training facilities at all levels;
(*b*) Introducing/strengthening refresher training courses, including fellowships and study tours, to update skills and technological know-how and improve productivity;
(*c*) Strengthening capability for research, planning, economic analysis, periodical evaluations and evaluation, relevant to improved utilization of forest resources;
(*d*) Promoting efficiency and capability of private and cooperative sectors through provision of facilities and incentives.

Capacity-building

Capacity-building, including strengthening of existing capacity, is implicit in the programme activities. Improving administration, policy and plans, national institutions, human resources, research and scientific capabilities, technology development, and periodical evaluations and evaluation are important components of capacity-building.

Establishing and/or Strengthening Capacities for the Planning, Assessment and Systematic Observations of Forests and Related Programmes, Projects and Activities, Including Commercial Trade and Processes

Basis for Action

Assessment and systematic observations are essential components of long-term planning, for evaluating effects, quantitatively and qualitatively, and for rectifying inadequacies. This mechanism, however, is one of the often neglected aspects of forest resources, management, conservation and development. In many cases, even the basic information related to the area and type of forests, existing potential and volume of harvest is lacking. In many developing countries, there is a lack of structures and mechanisms to carry out these functions. There is an urgent need to rectify this situation for a better understanding of the role and importance of forests and to realistically plan for their effective conservation, management, regeneration, and sustainable development.

Objectives

The objectives of this programme area are as follows:

(*a*) To strengthen or establish systems for the assessment and systematic observations of forests and forest lands with a view to assessing the impacts of programmes, projects and activities on the quality and extent of forest resources, land available for afforestation, and land tenure, and to integrate the systems in a continuing process of research and in-depth analysis, while ensuring necessary modifications and improvements for planning and decision-making. Specific emphasis should be given to the participation of rural people in these processes;

(*b*) To provide economists, planners, decision makers and local communities with sound and adequate updated information on forests and forest land resources.

Activities

Management-related Activities

Governments and institutions, in collaboration, where necessary, with appropriate international agencies and organizations, universities and non-governmental organizations, should undertake assessments and systematic observations of forests and related programmes and processes with a view to their continuous improvement. This should be linked to related activities of research and management and, wherever possible, be built upon existing systems. Major activities to be considered are:

(*a*) Assessing and carrying out systematic observations of the quantitative and qualitative situation and changes of forest cover and forest resources endowments, including land classification, land use and updates of its status, at the appropriate national level, and linking this activity, as appropriate, with planning as a basis for policy and programme formulation;
(*b*) Establishing national assessment and systematic observation systems and evaluation of programmes and processes, including establishment of definitions, standards, norms and intercalibration methods, and the capability for initiating corrective actions as well as improving the formulation and implementation of programmes and projects;
(*c*) Making estimates of impacts of activities affecting forestry developments and conservation proposals, in terms of key variables such as developmental goals, benefits and costs, contributions of forests to other sectors, community welfare, environmental conditions and biological diversity and their impacts at the local, regional and global levels, where appropriate, to assess the changing technological and financial needs of countries;

(*d*) Developing national systems of forest resource assessment and valuation, including necessary research and data analysis, which account for, where possible, the full range of wood and nonwood forest products and services, and incorporating results in plans and strategies and, where feasible, in national systems of accounts and planning;
(*e*) Establishing necessary intersectoral and programme linkages, including improved access to information, in order to support a holistic approach to planning and programming.

Data and Information

Reliable data and information are vital to this programme area. National Governments, in collaboration, where necessary, with relevant international organizations, should, as appropriate, undertake to improve data and information continuously and to ensure its exchange. Major activities to be considered are as follows:

(*a*) Collecting, consolidating and exchanging existing information and establishing baseline information on aspects relevant to this programme area;
(*b*) Harmonizing the methodologies for programmes involving data and information activities to ensure accuracy and consistency;
(*c*) Undertaking special surveys on, for example, land capability and suitability for afforestation action;
(*d*) Enhancing research support and improving access to and exchange of research results.

International and Regional Cooperation and Coordination

The international community should extend to the Governments concerned necessary technical and financial support for implementing this programme area, including consideration of the following activities:

(*a*) Establishing conceptual framework and formulating acceptable criteria, norms and definitions for systematic observations and assessment of forest resources;

(*b*) Establishing and strengthening national institutional coordination mechanisms for forest assessment and systematic observation activities;
(*c*) Strengthening existing regional and global networks for the exchange of relevant information;
(*d*) Strengthening the capacity and ability and improving the performance of existing international organizations, such as the Consultative Group on International Agricultural Research (CGIAR), FAO, ITTO, UNEP, UNESCO and UNIDO, to provide technical support and guidance in this programme area.

Means of Implementation

Financial and Cost Evaluation

The secretariat of the Conference has estimated the average total annual cost (1993-2000) of implementing the activities of this programme to be about $750 million, including about $230 million from the international community on grant or concessional terms. These are indicative and order-ofmagnitude estimates only and have not been reviewed by Governments. Actual costs and financial terms, including any that are non-concessional, will depend upon, *inter alia*, the specific strategies and programmes Governments decide upon for implementation.

Accelerating development consists of implementing the management-related and data/information activities cited above. Activities related to global environmental issues are those that will contribute to global information for assessing/evaluating/addressing environmental issues on a worldwide basis. Strengthening the capacity of international institutions consists of enhancing the technical staff and the executing capacity of several international organizations in order to meet the requirements of countries.

Scientific and Technological Means

Assessment and systematic observation activities involve major research efforts, statistical modelling and technological innovation. These have been internalized into the management-related activities. The activities in turn will improve the technological and scientific content of assessment and periodical evaluations. Some of the specific scientific and technological components included under these activities are:

(*a*) Developing technical, ecological and economic methods and models related to periodical evaluations and evaluation;
(*b*) Developing data systems, data processing and statistical modelling;
(*c*) Remote sensing and ground surveys;
(*d*) Developing geographic information systems;
(*e*) Assessing and improving technology.

These are to be linked and harmonized with similar activities and components in the other programme areas.

Human Resource Development

The programme activities foresee the need and include provision for human resource development in terms of specialization (*e.g.*, the use of remote-sensing, mapping and statistical modelling), training, technology transfer, fellowships and field demonstrations.

Capacity-building

National Governments, in collaboration with appropriate international organizations and institutions, should develop the necessary capacity for implementing this programme area. This should be harmonized with capacity-building for other programme areas. Capacity-building should cover such aspects as policies, public administration, national-level institutions, human resource and skill development, research capability, technology development, information systems, programme evaluation, intersectoral coordination and international cooperation.

Funding of International and Regional Cooperation

The secretariat of the Conference has estimated the average total annual cost (1993-2000) of implementing the activities of this

programme to be about $750 million, including about $530 million from the international community on grant or concessional terms. These are indicative and order-ofmagnitude estimates only and have not been reviewed by Governments. Actual costs and financial terms, including any that are non-concessional, will depend upon, *inter alia,* the specific strategies and programmes Governments decide upon for implementation.

MANAGING FRAGILE ECOSYSTEMS: COMBATING DESERTIFICATION AND DROUGHT

Fragile ecosystems are important ecosystems, with unique features and resources. Fragile ecosystems include deserts, semi-arid lands, mountains, wetlands, small islands and certain coastal areas. Most of these ecosystems are regional in scope, as they transcend national boundaries. This chapter addresses land resource issues in deserts, as well as arid, semi-arid and dry sub-humid areas.

Desertification is land degradation in arid, semi-arid and dry sub-humid areas resulting from various factors, including climatic variations and human activities. Desertification affects about one sixth of the world's population, 70 per cent of all drylands, amounting to 3.6 billion hectares, and one quarter of the total land area of the world. The most obvious impact of desertification, in addition to widespread poverty, is the degradation of 3.3 billion hectares of the total area of rangeland, constituting 73 per cent of the rangeland with a low potential for human and animal carrying capacity; decline in soil fertility and soil structure on about 47 per cent of the dryland areas constituting marginal rainfed cropland; and the degradation of irrigated cropland, amounting to 30 per cent of the dryland areas with a high population density and agricultural potential.

The priority in combating desertification should be the implementation of preventive measures for lands that are not yet degraded, or which are only slightly degraded. However, the severely degraded areas should not be neglected. In combating desertification and drought, the participation of local communities, rural organizations, national Governments, non-governmental organizations and international and regional organizations is essential.

The following programme areas are included in this chapter:

(*a*) Strengthening the knowledge base and developing information and monitoring systems for regions prone to desertification and drought, including the economic and social aspects of these ecosystems;
(*b*) Combating land degradation through, *inter alia*, intensified soil conservation, afforestation and reforestation activities;
(*c*) Developing and strengthening integrated development programmes for the eradication of poverty and promotion of alternative livelihood systems in areas prone to desertification;
(*d*) Developing comprehensive anti-desertification programmes and integrating them into national development plans and national environmental planning;
(*e*) Developing comprehensive drought preparedness and drought-relief schemes, including self-help arrangements, for drought-prone areas and designing programmes to cope with environmental refugees;
(*f*) Encouraging and promoting popular participation and environmental education, focusing on desertification control and management of the effects of drought.

Strengthening the Knowledge Base and Developing Information and Monitoring Systems for Regions Prone to Desertification and Drought, Including the Economic and Social Aspects of these Ecosystems

Basis for Action 12.5. The global assessments of the status and rate of desertification conducted by the United Nations Environment Programme (UNEP) in 1977, 1984 and 1991 have revealed insufficient basic knowledge of desertification processes. Adequate world-wide systematic observation systems are helpful for the development and implementation of effective anti-desertification programmes. The capacity of existing international, regional and national institutions, particularly in developing countries, to generate and exchange relevant information is limited. An integrated and coordinated information and systematic observation system based on appropriate technology and embracing global, regional, national and local levels is essential for understanding the dynamics of desertification and drought processes. It is also important for developing adequate measures to deal with desertification and drought and improving socio-economic conditions. Objectives

The objectives of this programme area are:

(*a*) To promote the establishment and/or strengthening of national environmental information coordination centres that will act as focal points within Governments for sectoral ministries and provide the necessary standardization and back-up services; to ensure also that national environmental information systems on desertification and drought are linked together through a network at subregional, regional and interregional levels;

(*b*) To strengthen regional and global systematic observation networks linked to the development of national systems for the observation of land degradation and desertification caused both by climate fluctuations and by human impact, and to identify priority areas for action;

(*c*) To establish a permanent system at both national and international levels for monitoring desertification and land degradation with the aim of improving living conditions in the affected areas.

Activities

Management-related Activities

Governments at the appropriate level, with the support of the relevant international and regional organizations, should:

(*a*) Establish and/or strengthen environmental information systems at the national level;

(*b*) Strengthen national, state/provincial and local assessment and ensure cooperation/networking between existing environmental information and monitoring systems, such as Earthwatch and the Sahara and Sahel Observatory;

(*c*) Strengthen the capacity of national institutions to analyse environmental data so that ecological change can be monitored and environmental information obtained on a continuing basis at the national level.

Data and Information

Governments at the appropriate level, with the support of the relevant international and regional organizations, should:

(*a*) Review and study the means for measuring the ecological, economic and social consequences of desertification and land degradation and introduce the results of these studies internationally into desertification and land degradation assessment practices;

(*b*) Review and study the interactions between the socio-economic impacts of climate, drought and desertification and utilize the results of these studies to secure concrete action.

Governments at the appropriate level, with the support of the relevant international and regional organizations, should:

(*a*) Support the integrated data collection and research work of programmes related to desertification and drought problems;

(*b*) Support national, regional and global programmes for integrated data collection and research networks carrying out assessment of soil and land degradation;

(*c*) Strengthen national and regional meteorological and hydrological networks and monitoring systems to ensure adequate collection of basic information and communication among national, regional and international centres.

International and Regional Cooperation and Coordination

Governments at the appropriate level, with the support of the relevant international and regional organizations, should:

(*a*) Strengthen regional programmes and international cooperation, such as the Permanent Inter-State Committee on Drought Control in the Sahel (CILSS), the Intergovernmental Authority for Drought and Development (IGADD), the Southern African Development Coordination Conference (SADCC), the Arab Maghreb Union and other regional organizations, as well as such organizations as the Sahara and Sahel Observatory;

(*b*) Establish and/or develop a comprehensive desertification, land degradation and human condition database component that incorporates both physical and socio-economic parameters. This should be based on existing and, where necessary, additional facilities, such as those of Earthwatch and other information systems of international, regional and national institutions strengthened for this purpose;

(*c*) Determine benchmarks and define indicators of progress that facilitate the work of local and regional organizations in tracking progress in the fight for anti-desertification. Particular attention should be paid to indicators of local participation.

Means of Implementation

Financing and Cost Evaluation

The Conference secretariat has estimated the average total annual cost (1993-2000) of implementing the activities of this programme to be about $350 million, including about $175 million from the international community on grant or concessional terms. These are indicative and order-of-magnitude estimates only and have not been reviewed by Governments. Actual costs and financial terms, including any that are non-concessional, will depend upon, *inter alia*, the specific strategies and programmes Governments decide upon for implementation.

Scientific and Technological Means

Governments at the appropriate level, with the support of the relevant international and regional organizations working on the issue of desertification and drought, should:

(*a*) Undertake and update existing inventories of natural resources, such as energy, water, soil, minerals, plant and animal access to food, as well as other resources, such as housing, employment, health, education and demographic distribution in time and space;
(*b*) Develop integrated information systems for environmental monitoring, accounting and impact assessment;
(*c*) International bodies should cooperate with national Governments to facilitate the acquisition and development of appropriate technology for monitoring and combating drought and desertification.

Human Resource Development

Governments at the appropriate level, with the support of the relevant international and regional organizations working on the issue of desertification and drought, should develop the technical and professional skills of people engaged in monitoring and assessing the issue of desertification and drought.

Capacity-building

Governments at the appropriate level, with the support of the relevant international and regional organizations working on the issue of desertification and drought, should:

(*a*) Strengthen national and local institutions by providing adequate staff equipment and finance for assessing desertification;
(*b*) Promote the involvement of the local population, particularly women and youth, in the collection and utilization of environmental information through education and awareness building.

Combating Land Degradation through, *inter alia,* Intensified Soil Conservation, Afforestation and Reforestation Activities

Basis for Action

Desertification affects about 3.6 billion hectares, which is about 70 per cent of the total area of the world's drylands or nearly one quarter of the global land area. In combating desertification on rangeland, rainfed cropland and irrigated land, preventative measures should be launched in areas which are not yet affected or are only slightly affected by desertification; corrective measures should be implemented to sustain the productivity of moderately desertified land; and rehabilitative measures should be taken to recover severely or very severely desertified drylands.

An increasing vegetation cover would promote and stabilize the hydrological balance in the dryland areas and maintain land quality and land productivity. Prevention of not yet degraded land and application of corrective measures and rehabilitation of moderate and severely degraded drylands, including areas affected by sand dune movements, through the introduction of environmentally sound, socially acceptable, fair and economically feasible land-use systems. This will enhance the land carrying capacity and maintenance of biotic resources in fragile ecosystems.

Objectives

The objectives of this programme area are:

(*a*) As regards areas not yet affected or only slightly affected by desertification, to ensure appropriate management of existing natural formations (including forests) for the conservation of biodiversity, watershed protection, sustainability of their production and agricultural development, and other purposes, with the full participation of indigenous people;

(*b*) To rehabilitate moderately to severely desertified drylands for productive utilization and sustain their productivity for agropastoral/agroforestry development through, *inter alia*, soil and water conservation;

(*c*) To increase the vegetation cover and support management of biotic resources in regions affected or prone to desertification and drought, notably through such activities as afforestation/reforestation, agroforestry, community forestry and vegetation retention schemes;

(*d*) To improve management of forest resources, including woodfuel, and to reduce woodfuel consumption through more efficient utilization, conservation and the enhancement, development and use of other sources of energy, including alternative sources of energy.

Activities

Management-related Activities

Governments at the appropriate level, and with the support of the relevant international and regional organizations, should:

(*a*) Implement urgent direct preventive measures in drylands that are vulnerable but not yet affected, or only slightly desertified drylands, by introducing (*i*) improved land-use policies and practices for more sustainable land productivity; (*ii*) appropriate, environmentally sound and economically feasible agricultural and pastoral technologies; and (*iii*) improved management of soil and water resources;

(*b*) Carry out accelerated afforestation and reforestation programmes, using droughtresistant, fast-growing species, in particular native ones, including legumes and other species, combined with community-based agroforestry schemes. In this regard, creation of large-scale reforestation and afforestation schemes, particularly through the establishment of green belts, should be considered, bearing in mind the multiple benefits of such measures;

(*c*) Implement urgent direct corrective measures in moderately to severely desertified drylands, in addition to the measures listed in paragraph 19(*a*) above, with a view to restoring and sustaining their productivity;

(*d*) Promote improved land/water/crop-management systems, making it possible to combat salinization in existing irrigated croplands; and to stabilize rainfed croplands and introduce improved soil/crop-management systems into land-use practice;

(*e*) Promote participatory management of natural resources, including rangeland, to meet both the needs of rural populations and conservation purposes, based on innovative or adapted indigenous technologies;

(*f*) Promote *in situ* protection and conservation of special ecological areas through legislation and other means for the purpose of combating desertification while ensuring the protection of biodiversity;

(*g*) Promote and encourage investment in forestry development in drylands through various incentives, including legislative measures;

(*h*) Promote the development and use of sources of energy which will lessen pressure on ligneous resources, including alternative sources of energy and improved stoves.

Data and Information

Governments at the appropriate level, with the support of the relevant international and regional organizations, should:

(*a*) Develop land-use models based on local practices for the improvement of such practices, with a focus on preventing land degradation. The models should give a better understanding of the variety of natural and human-induced factors that may contribute to desertification. Models should incorporate the interaction of both new and traditional practices to prevent land degradation and reflect the resilience of the whole ecological and social system;

(*b*) Develop, test and introduce, with due regard to environmental security considerations, drought resistant, fast-growing and productive plant species appropriate to the environment of the regions concerned.

International and Regional Cooperation and Coordination

The appropriate United Nations agencies, international and regional organizations, non-governmental organizations and bilateral agencies should:

(*a*) Coordinate their roles in combating land degradation and promoting reforestation, agroforestry and land-management systems in affected countries;

(*b*) Support regional and subregional activities in technology development and dissemination, training and programme implementation to arrest dryland degradation.

The National Governments concerned, the appropriate United Nations agencies and bilateral agencies should strengthen the coordinating role in dryland degradation of subregional intergovernmental organizations set up to cover these activities, such as CILSS, IGADD, SADCC and the Arab Maghreb Union.

Means of Implementation

Financing and Cost Evaluation

The Conference secretariat has estimated the average total annual cost (1993-2000) of implementing the activities of this programme to be about $6 billion, including about $3 billion from the international community on grant or concessional terms. These are indicative and order-ofmagnitude estimates only and have not been reviewed by Governments. Actual costs and financial terms, including any that are non-concessional, will depend upon, *inter alia,* the specific strategies and programmes Governments decide upon for implementation.

Scientific and Technological Means

Governments at the appropriate level and local communities, with the support of the relevant international and regional organizations, should:

(*a*) Integrate indigenous knowledge related to forests, forest lands, rangeland and natural vegetation into research activities on desertification and drought;

(*b*) Promote integrated research programmes on the protection, restoration and conservation of water and land resources and land-use management based on traditional approaches, where feasible.

Human Resource Development

Governments at the appropriate level and local communities, with the support of the relevant international and regional organizations, should:

(*a*) Establish mechanisms to ensure that land users, particularly women, are the main actors in implementing improved land use, including agroforestry systems, in combating land degradation;

(*b*) Promote efficient extension-service facilities in areas prone to desertification and drought, particularly for training farmers and pastoralists in the improved management of land and water resources in drylands.

Capacity-building

Governments at the appropriate level and local communities, with the support of the relevant international and regional organizations, should:

(*a*) Develop and adopt, through appropriate national legislation, and introduce institutionally, new and environmentally sound development-oriented land-use policies;

(*b*) Support community-based people's organizations, especially farmers and pastoralists.

Developing and Strengthening Integrated Development Programmes for the Eradication of Poverty and Promotion of Alternative Livelihood Systems in Areas Prone to Desertification

Basis for Action

In areas prone to desertification and drought, current livelihood and resource-use systems are not able to maintain living standards. In most of the arid and semi-arid areas, the traditional livelihood systems based on agropastoral systems are often inadequate and unsustainable, particularly in view of the effects of drought and increasing demographic pressure. Poverty is a major factor in accelerating the rate of degradation and desertification. Action is therefore needed to rehabilitate and improve the agropastoral systems for sustainable management of rangelands, as well as alternative livelihood systems.

Objectives

The objectives of this programme area are:

(*a*) To create the capacity of village communities and pastoral groups to take charge of their development and the management of their land resources on a socially equitable and ecologically sound basis;
(*b*) To improve production systems in order to achieve greater productivity within approved programmes for conservation of national resources and in the framework of an integrated approach to rural development;
(*c*) To provide opportunities for alternative livelihoods as a basis for reducing pressure on land resources while at the same time providing additional sources of income, particularly for rural populations, thereby improving their standard of living.

Activities

Management-related Activities

Governments at the appropriate level, with the support of the relevant international and regional organizations, should:

(*a*) Adopt policies at the national level regarding a decentralized approach to land-resource management, delegating responsibility to rural organizations;
(*b*) Create or strengthen rural organizations in charge of village and pastoral land management;
(*c*) Establish and develop local, national and intersectoral mechanisms to handle environmental and develop mental consequences of land tenure expressed in terms of land use and land ownership. Particular attention should be given to protecting the property rights of women and pastoral and nomadic groups living in rural areas;
(*d*) Create or strengthen village associations focused on economic activities of common pastoral interest (market gardening, transformation of agricultural products, livestock, herding, etc.);

(*e*) Promote rural credit and mobilization of rural savings through the establishment of rural banking systems;

(*f*) Develop infrastructure, as well as local production and marketing capacity, by involving the local people to promote alternative livelihood systems and alleviate poverty;

(*g*) Establish a revolving fund for credit to rural entrepreneurs and local groups to facilitate the establishment of cottage industries/business ventures and credit for input to agropastoral activities.

Data and Information

Governments at the appropriate level, with the support of the relevant international and regional organizations, should:

(*a*) Conduct socio-economic baseline studies in order to have a good understanding of the situation in the programme area regarding, particularly, resource and land tenure issues, traditional land-management practices and characteristics of production systems;

(*b*) Conduct inventory of natural resources (soil, water and vegetation) and their state of degradation, based primarily on the knowledge of the local population (*e.g.*, rapid rural appraisal);

(*c*) Disseminate information on technical packages adapted to the social, economic and ecological conditions of each;

(*d*) Promote exchange and sharing of information concerning the development of alternative livelihoods with other agro-ecological regions.

International and Regional Cooperation and Coordination

Governments at the appropriate level, and with the support of the relevant international and regional organizations, should:

(*a*) Promote cooperation and exchange of information among the arid and semi-arid land research institutions concerning techniques and technologies to improve land and labour productivity, as well as viable production systems;

(*b*) Coordinate and harmonize the implementation of programmes and projects funded by the international organization communities and non-governmental organizations that are directed towards the alleviation of poverty and promotion of an alternative livelihood system.

Means of Implementation

Financing and Cost Evaluation

The Conference secretariat has estimated the costs for this programme area and Promoting sustainable agriculture and rural development.

Scientific and Technological Means

Governments at the appropriate level, and with the support of the relevant international and regional organizations, should:

(*a*) Undertake applied research in land use with the support of local research institutions;

(*b*) Facilitate regular national, regional and interregional communication on and exchange of information and experience between extension officers and researchers;

(*c*) Support and encourage the introduction and use of technologies for the generation of alternative sources of incomes.

Human Resource Development

Governments at the appropriate level, with the support of the relevant international and regional organizations, should:

(*a*) Train members of rural organizations in management skills and train agropastoralists in such special techniques as soil and water conservation, water harvesting, agroforestry and small-scale irrigation;

(*b*) Train extension agents and officers in the participatory approach to integrated land management.

Capacity-building

Governments at the appropriate level, with the support of the relevant international and regional organizations, should establish and maintain mechanisms to ensure the integration into sectoral and national development plans and programmes of strategies for poverty alleviation among the inhabitants of lands prone to desertification.

Developing Comprehensive Anti-desertification Programmes and Integrating them into National Development Plans and National Environmental Planning

Basis for Action

In a number of developing countries affected by desertification, the natural resource base is the main resource upon which the development process must rely. The social systems interacting with land resources make the problem much more complex, requiring an integrated approach to the planning and management of land resources. Action plans to combat desertification and drought should include management aspects of the environment and development, thus conforming with the approach of integrating national development plans and national environmental action plans.

Objectives

The objectives of this programme area are:

- (*a*) To strengthen national institutional capabilities to develop appropriate anti-desertification programmes and to integrate them into national development planning;
- (*b*) To develop and integrate strategic planning frameworks for the development, protection and management of natural resources in dryland areas into national development plans, including national plans to combat desertification, and environmental action plans in countries most prone to desertification;

(*c*) To initiate a long-term process for implementing and monitoring strategies related to natural resources management;

(*d*) To strengthen regional and international cooperation for combating desertification through, *inter alia*, the adoption of legal and other instruments.

Activities

Management-related Activities

Governments at the appropriate level, and with the support of the relevant international and regional organizations, should:

(*a*) Establish or strengthen, national and local anti-desertification authorities within government and local executive bodies, as well as local committees/associations of land users, in all rural communities affected, with a view to organizing working cooperation between all actors concerned, from the grass-roots level (farmers and pastoralists) to the higher levels of government;

(*b*) Develop national plans of action to combat desertification and as appropriate, make them integral parts of national development plans and national environmental action plans;

(*c*) Implement policies directed towards improving land use, managing common lands appropriately, providing incentives to small farmers and pastoralists, involving women and encouraging private investment in the development of drylands;

(*d*) Ensure coordination among ministries and institutions working on anti-desertification programmes at national and local levels.

Data and Information

Governments at the appropriate level, and with the support of the relevant international and regional organizations, should promote information exchange and cooperation with respect to national planning and programming among affected countries, *inter alia*, through networking.

International and Regional Cooperation and Coordination

The relevant international organizations, multilateral financial institutions, non-governmental organizations and bilateral agencies should strengthen their cooperation in assisting with the preparation of desertification control programmes and their integration into national planning strategies, with the establishment of national coordinating and systematic observation mechanisms and with the regional and global networking of these plans and mechanisms.

The General Assembly, at its forty-seventh session, should be requested to establish, under the aegis of the General Assembly, an intergovernmental negotiating committee for the elaboration of an international convention to combat desertification in those countries experiencing serious drought and/or desertification, particularly in Africa, with a view to finalizing such a convention by June 1994.

Means of Implementation

Financing and Cost Evaluation

The Conference secretariat has estimated the average total annual cost (1993-2000) of implementing the activities of this programme to be about $180 million, including about $90 million from the international community on grant or concessional terms. These are indicative and order-ofmagnitude estimates only and have not been reviewed by Governments. Actual costs and financial terms, including any that are non-concessional, will depend upon, *inter alia*, the specific strategies and programmes Governments decide upon for implementation.

Scientific and Technological Means

Governments at the appropriate level, with the support of the relevant international and regional organizations, should:

(*a*) Develop and introduce appropriate improved sustainable agricultural and pastoral technologies that are socially and environmentally acceptable and economically feasible;

(*b*) Undertake applied study on the integration of environmental and developmental activities into national development plans.

Human Resource Development

Governments at the appropriate level, with the support of the relevant international and regional organizations, should undertake nationwide major anti-desertification awareness/training campaigns within countries affected through existing national mass media facilities, educational networks and newly created or strengthened extension services. This should ensure people's access to knowledge of desertification and drought and to national plans of action to combat desertification.

Capacity-building

Governments at the appropriate level, with the support of the relevant international and regional organizations, should establish and maintain mechanisms to ensure coordination of sectoral ministries and institutions, including local-level institutions and appropriate non-governmental organizations, in integrating anti-desertification programmes into national development plans and national environmental action plans.

Developing Comprehensive Drought Preparedness and Drought-relief Schemes, Including Self-help Arrangements, for Drought-prone Areas and Designing Programmes to Cope with Environmental Refugees

Basis for Action

Drought, in differing degrees of frequency and severity, is a recurring phenomenon throughout much of the developing world, especially Africa. Apart from the human toll—an estimated 3 million people died in the mid-1980s because of drought in sub-Saharan Africa—the economic costs of drought-related disasters are also high in terms of lost production, misused inputs and diversion of development resources.

Early-warning systems to forecast drought will make possible the implementation of drought—preparedness schemes. Integrated packages at the farm and watershed level, such as alternative cropping strategies, soil and water conservation and promotion of water harvesting techniques, could enhance the capacity of land to cope with drought and provide basic necessities, thereby minimizing the number of environmental refugees and the need for emergency drought relief. At the same time, contingency arrangements for relief are needed for periods of acute scarcity.

Objectives

The objectives of this programme area are:

(*a*) To develop national strategies for drought preparedness in both the short and long-term, aimed at reducing the vulnerability of production systems to drought;
(*b*) To strengthen the flow of early-warning information to decision makers and land users to enable nations to implement strategies for drought intervention;
(*c*) To develop and integrate drought-relief schemes and means of coping with environmental refugees into national and regional development planning.

Activities

Management-related Activities

In drought-prone areas, Governments at the appropriate level, with the support of the relevant international and regional organizations, should:

(*a*) Design strategies to deal with national food deficiencies in periods of production shortfall. These strategies should deal with issues of storage and stocks, imports, port facilities, food storage, transport and distribution;
(*b*) Improve national and regional capacity for agrometeorology and contingency crop planning. Agrometeorology links the frequency, content and regional coverage of weather

forecasts with the requirements of crop planning and agricultural extension;

(*c*) Prepare rural projects for providing short-term rural employment to drought-affected households. The loss of income and entitlement to food is a common source of distress in times of drought. Rural works help to generate the income required to buy food for poor households;

(*d*) Establish contingency arrangements, where necessary, for food and fodder distribution and water supply;

(*e*) Establish budgetary mechanisms for providing, at short notice, resources for drought relief;

(*f*) Establish safety nets for the most vulnerable households.

Data and Information

Governments of affected countries, at the appropriate level, with the support of the relevant international and regional organizations, should:

(*a*) Implement research on seasonal forecasts to improve contingency planning and relief operations and allow preventive measures to be taken at the farm level, such as the selection of appropriate varieties and farming practices, in times of drought;

(*b*) Support applied research on ways of reducing water loss from soils, on ways of increasing the water absorption capacities of soils and on water harvesting techniques in drought-prone areas;

(*c*) Strengthen national early-warning systems, with particular emphasis on the area of riskmapping, remote-sensing, agrometeorological modelling, integrated multidisciplinary crop-forecasting techniques and computerized food supply/demand analysis.

International and Regional Cooperation and Coordination

Governments at the appropriate level, with the support of the relevant international and regional organizations, should:

(*a*) Establish a system of stand-by capacities in terms of foodstock, logistical support, personnel and finance for a speedy international response to drought-related emergencies;

(*b*) Support programmes of the World Meteorological Organization (WMO) on agrohydrology and agrometeorology, the Programme of the Regional Training Centre for Agrometeorology and Operational Hydrology and their Applications (AGRHYMET), drought-monitoring centres and the African Centre of Meteorological Applications for Development (ACMAD), as well as the efforts of the Permanent Inter-State Committee on Drought Control in the Sahel (CILSS) and the Intergovernmental Authority for Drought and Development (IGADD);

(*c*) Support FAO programmes and other programmes for the development of national early—warning systems and food security assistance schemes;

(*d*) Strengthen and expand the scope of existing regional programmes and the activities of appropriate United Nations organs and organizations, such as the World Food Programme (WFP), the Office of the United Nations Disaster Relief Coordinator (UNDRO) and the United Nations Sudano-Sahelian Office as well as of non-governmental organizations, aimed at mitigating the effects of drought and emergencies.

Means of Implementation

Financing and Cost Evaluation

The Conference secretariat has estimated the average total annual cost (1993-2000) of implementing the activities of this programme to be about $1.2 billion, including about $1.1 billion from the international community on grant or concessional terms. These are indicative and order-ofmagnitude estimates only and have not been reviewed by Governments. Actual costs and financial terms, including any that are non-concessional, will depend upon, *inter alia*, the specific strategies and programmes Governments decide upon for implementation.

Scientific and Technological Means

Governments at the appropriate level and drought-prone communities, with the support of the relevant international and regional organizations, should:

(*a*) Use traditional mechanisms to cope with hunger as a means of channelling relief and development assistance;

(*b*) Strengthen and develop national, regional and local interdisciplinary research and training capabilities for drought-prevention strategies.

Human Resource Development

Governments at the appropriate level, with the support of the relevant international and regional organizations, should:

(*a*) Promote the training of decision makers and land users in the effective utilization of information from early-warning systems;

(*b*) Strengthen research and national training capabilities to assess the impact of drought and to develop methodologies to forecast drought.

Capacity-building

Governments at the appropriate level, with the support of the relevant international and regional organizations, should:

(*a*) Improve and maintain mechanisms with adequate staff, equipment and finances for monitoring drought parameters to take preventive measures at regional, national and local levels;

(*b*) Establish interministerial linkages and coordinating units for drought monitoring, impact assessment and management of drought-relief schemes.

Encouraging and Promoting Popular Participation and Environmental Education, Focusing on Desertification Control and Management of the Effects of Drought

Basis for Action

The experience to date on the successes and failures of programmes and projects points to the need for popular support to sustain activities related to desertification and drought control. But it is necessary to go beyond the theoretical ideal of popular participation and to focus on obtaining actual active popular involvement, rooted in the concept of partnership. This implies the sharing of responsibilities and the mutual involvement of all parties. In this context, this programme area should be considered an essential supporting component of all desertification-control and drought-related activities.

Objectives

The objectives of this programme area are:

(*a*) To develop and increase public awareness and knowledge concerning desertification and drought, including the integration of environmental education in the curriculum of primary and secondary schools;

(*b*) To establish and promote true partnership between government authorities, at both the national and local levels, other executing agencies, non-governmental organizations and land users stricken by drought and desertification, giving land users a responsible role in the planning and execution processes in order to benefit fully from development projects;

(*c*) To ensure that the partners understand one another's needs, objectives and points of view by providing a variety of means such as training, public awareness and open dialogue;

(*d*) To support local communities in their own efforts in combating desertification, and to draw on the knowledge and experience of the populations concerned, ensuring the full participation of women and indigenous populations.

Activities

Management-related Activities

Governments at the appropriate level, with the support of the relevant international and regional organizations, should:

(*a*) Adopt policies and establish administrative structures for more decentralized decision-making and implementation;
(*b*) Establish and utilize mechanisms for the consultation and involvement of land users and for enhancing capability at the grass-roots level to identify and/or contribute to the identification and planning of action;
(*c*) Define specific programme/project objectives in cooperation with local communities; design local management plans to include such measures of progress, thereby providing a means of altering project design or changing management practices, as appropriate;
(*d*) Introduce legislative, institutional/organizational and financial measures to secure user involvement and access to land resources;
(*e*) Establish and/or expand favourable conditions for the provision of services, such as credit facilities and marketing outlets for rural populations;
(*f*) Develop training programmes to increase the level of education and participation of people, particularly women and indigenous groups, through, *inter alia*, literacy and the development of technical skills;
(*g*) Create rural banking systems to facilitate access to credit for rural populations, particularly women and indigenous groups, and to promote rural savings;
(*h*) Adopt appropriate policies to stimulate private and public investment.

Data and Information

Governments at the appropriate level, with the support of the relevant international and regional organizations, should:

(*a*) Review, develop and disseminate gender-disaggregated information, skills and knowhow at all levels on ways of organizing and promoting popular participation;
(*b*) Accelerate the development of technological know-how, focusing on appropriate and intermediate technology;
(*c*) Disseminate knowledge about applied research results on soil and water issues, appropriate species, agricultural techniques and technological know-how.

International and Regional Cooperation and Coordination

Governments at the appropriate level, and with the support of the relevant international and regional organizations, should:

(*a*) Develop programmes of support to regional organizations such as CILSS, IGADD, SADCC and the Arab Maghreb Union and other intergovernmental organizations in Africa and other parts of the world, to strengthen outreach programmes and increase the participation of non-governmental organizations together with rural populations;

(*b*) Develop mechanisms for facilitating cooperation in technology and promote such cooperation as an element of all external assistance and activities related to technical assistance projects in the public or private sector;

(*c*) Promote collaboration among different actors in environment and development programmes;

(*d*) Encourage the emergence of representative organizational structures to foster and sustain interorganizational cooperation.

Means of Implementation

Financing and Cost Evaluation

The Conference secretariat has estimated the average total annual cost (1993-2000) of implementing the activities of this programme to be about $1.0 billion, including about $500 million from the international community on grant or concessional terms. These are indicative and order-ofmagnitude estimates only and have not been reviewed by Governments. Actual costs and financial terms, including any that are non-concessional, will depend upon, *inter alia*, the specific strategies and programmes Governments decide upon for implementation.

Scientific and Technological Means

Governments at the appropriate level, and with the support of the relevant international and regional organizations, should promote the development of indigenous know-how and technology transfer.

Human Resource Development

Governments, at the appropriate level, and with the support of the relevant international and regional organizations, should:

(*a*) Support and/or strengthen institutions involved in public education, including the local media, schools and community groups;
(*b*) Increase the level of public education.

Capacity-building

Governments at the appropriate level, and with the support of the relevant international and regional organizations, should promote members of local rural organizations and train and appoint more extension officers working at the local level.

MANAGING FRAGILE ECOSYSTEMS: SUSTAINABLE MOUNTAIN DEVELOPMENT

Mountains are an important source of water, energy and biological diversity. Furthermore, they are a source of such key resources as minerals, forest products and agricultural products and of recreation. As a major ecosystem representing the complex and interrelated ecology of our planet, mountain environments are essential to the survival of the global ecosystem. Mountain ecosystems are, however, rapidly changing. They are susceptible to accelerated soil erosion, landslides and rapid loss of habitat and genetic diversity. On the human side, there is widespread poverty among mountain inhabitants and loss of indigenous knowledge. As a result, most global mountain areas are experiencing environmental degradation. Hence, the proper management of mountain resources and socio-economic development of the people deserves immediate action.

About 10 per cent of the world's population depends on mountain resources. A much larger percentage draws on other mountain resources, including and especially water. Mountains are a storehouse of biological diversity and endangered species.

Two programme areas are included in this chapter to further elaborate the problem of fragile ecosystems with regard to all mountains of the world. These are:

(*a*) Generating and strengthening knowledge about the ecology and sustainable development of mountain ecosystems;

(*b*) Promoting integrated watershed development and alternative livelihood opportunities.

Generating and Strengthening Knowledge about the Ecology and Sustainable Development of Mountain Ecosystems

Basis for Action

Mountains are highly vulnerable to human and natural ecological imbalance. Mountains are the areas most sensitive to all climatic changes in the atmosphere. Specific information on ecology, natural resource potential and socio-economic activities is essential. Mountain and hillside areas hold a rich variety of ecological systems. Because of their vertical dimensions, mountains create gradients of temperature, precipitation and insolation. A given mountain slope may include several climatic systems—such as tropical, subtropical, temperate and alpine—each of which represents a microcosm of a larger habitat diversity. There is, however, a lack of knowledge of mountain ecosystems. The creation of a global mountain database is therefore vital for launching programmes that contribute to the sustainable development of mountain ecosystems.

Objectives

The objectives of this programme area are:

(*a*) To undertake a survey of the different forms of soils, forest, water use, crop, plant and animal resources of mountain ecosystems, taking into account the work of existing international and regional organizations;

(*b*) To maintain and generate database and information systems to facilitate the integrated management and environmental assessment of mountain ecosystems, taking into account the work of existing international and regional organizations;

(*c*) To improve and build the existing land/water ecological knowledge base regarding technologies and agricultural and conservation practices in the mountain regions of the world, with the participation of local communities;
(*d*) To create and strengthen the communications network and information clearing-house for existing organizations concerned with mountain issues;
(*e*) To improve coordination of regional efforts to protect fragile mountain ecosystems through the consideration of appropriate mechanisms, including regional legal and other instruments;
(*f*) To generate information to establish databases and information systems to facilitate an evaluation of environmental risks and natural disasters in mountain ecosystems.

Activities

Management-related Activities

Governments at the appropriate level, with the support of the relevant international and regional organizations, should:

(*a*) Strengthen existing institutions or establish new ones at local, national and regional levels to generate a multidisciplinary land/water ecological knowledge base on mountain ecosystems;
(*b*) Promote national policies that would provide incentives to local people for the use and transfer of environment-friendly technologies and farming and conservation practices;
(*c*) Build up the knowledge base and understanding by creating mechanisms for cooperation and information exchange among national and regional institutions working on fragile ecosystems;
(*d*) Encourage policies that would provide incentives to farmers and local people to undertake conservation and regenerative measures;
(*e*) Diversify mountain economies, *inter alia*, by creating and/or strengthening tourism, in accordance with integrated management of mountain areas;

(*f*) Integrate all forest, rangeland and wildlife activities in such a way that specific mountain ecosystems are maintained;
(*g*) Establish appropriate natural reserves in representative species-rich sites and areas.

Data and Information

Governments at the appropriate level, with the support of the relevant international and regional organizations, should:

(*a*) Maintain and establish meteorological, hydrological and physical monitoring analysis and capabilities that would encompass the climatic diversity as well as water distribution of various mountain regions of the world;
(*b*) Build an inventory of different forms of soils, forests, water use, and crop, plant and animal genetic resources, giving priority to those under threat of extinction. Genetic resources should be protected *in situ* by maintaining and establishing protected areas and improving traditional farming and animal husbandry activities and establishing programmes for evaluating the potential value of the resources;
(*c*) Identify hazardous areas that are most vulnerable to erosion, floods, landslides, earthquakes, snow avalanches and other natural hazards;
(*d*) Identify mountain areas threatened by air pollution from neighbouring industrial and urban areas.

International and Regional Cooperation

National Governments and intergovernmental organizations should:

(*a*) Coordinate regional and international cooperation and facilitate an exchange of information and experience among the specialized agencies, the World Bank, IFAD and other international and regional organizations, national Governments, research institutions and non-governmental organizations working on mountain development;

(*b*) Encourage regional, national and international networking of people's initiatives and the activities of international, regional and local non-governmental organizations working on mountain development, such as the United Nations University (UNU), the Woodland Mountain Institutes (WMI), the International Center for Integrated Mountain Development (ICIMOD), the International Mountain Society (IMS), the African Mountain Association and the Andean Mountain Association, besides supporting those organizations in exchange of information and experience;

(*c*) Protect Fragile Mountain Ecosystem through the consideration of appropriate mechanisms including regional legal and other instruments.

Means of Implementation

Financing and Cost Evaluation

The Conference secretariat has estimated the average total annual cost (1993-2000) of implementing the activities of this programme to be about $50 million from the international community on grant or concessional terms. These are indicative and order-of-magnitude estimates only and have not been reviewed by Governments. Actual costs and financial terms, including any that are non-concessional, will depend upon, *inter alia*, the specific strategies and programmes Governments decide upon for implementation.

Scientific and Technological Means

Governments at the appropriate level, with the support of the relevant international and regional organizations, should strengthen scientific research and technological development programmes, including diffusion through national and regional institutions, particularly in meteorology, hydrology, forestry, soil sciences and plant sciences.

Human Resource Development

Governments at the appropriate level, and with the support of the relevant international and regional organizations, should:

(*a*) Launch training and extension programmes in environmentally appropriate technologies and practices that would be suitable to mountain ecosystems;

(*b*) Support higher education through fellowships and research grants for environmental studies in mountains and hill areas, particularly for candidates from indigenous mountain populations;

(*c*) Undertake environmental education for farmers, in particular for women, to help the rural population better understand the ecological issues regarding the sustainable development of mountain ecosystems.

Capacity-building

Governments at the appropriate level, with the support of the relevant international and regional organizations, should build up national and regional institutional bases that could carry out research, training and dissemination of information on the sustainable development of the economies of fragile ecosystems.

Promoting Integrated Watershed Development and Alternative Livelihood Opportunities

Basis for Action

Nearly half of the world's population is affected in various ways by mountain ecology and the degradation of watershed areas. About 10 per cent of the Earth's population lives in mountain areas with higher slopes, while about 40 per cent occupies the adjacent medium- and lower-watershed areas. There are serious problems of ecological deterioration in these watershed areas. For example, in the hillside areas of the Andean countries of South America a large portion of the farming population is now faced with a rapid deterioration of land resources. Similarly, the mountain and upland areas of the Himalayas, South-East Asia and East and Central Africa, which make vital contributions to agricultural production, are threatened by cultivation of marginal lands due to expanding population. In many areas this is accompanied by excessive livestock grazing, deforestation and loss of biomass cover.

Soil erosion can have a devastating impact on the vast numbers of rural people who depend on rainfed agriculture in the mountain and hillside areas. Poverty, unemployment, poor health and bad sanitation are widespread. Promoting integrated watershed development programmes through effective participation of local people is a key to preventing further ecological imbalance. An integrated approach is needed for conserving, upgrading and using the natural resource base of land, water, plant, animal and human resources. In addition, promoting alternative livelihood opportunities, particularly through development of employment schemes that increase the productive base, will have a significant role in improving the standard of living among the large rural population living in mountain ecosystems.

Objectives

The objectives of this programme area are:

(*a*) By the year 2000, to develop appropriate land-use planning and management for both arable and non-arable land in mountain-fed watershed areas to prevent soil erosion, increase biomass production and maintain the ecological balance;

(*b*) To promote income-generating activities, such as sustainable tourism, fisheries and environmentally sound mining, and to improve infrastructure and social services, in particular to protect the livelihoods of local communities and indigenous people;

(*c*) To develop technical and institutional arrangements for affected countries to mitigate the effects of natural disasters through hazard-prevention measures, risk zoning, earlywarning systems, evacuation plans and emergency supplies.

Activities

Management-related Activities

Governments at the appropriate level, with the support of the relevant international and regional organizations, should:

(*a*) Undertake measures to prevent soil erosion and promote erosion-control activities in all sectors;

(*b*) Establish task forces or watershed development committees, complementing existing institutions, to coordinate integrated services to support local initiatives in animal husbandry, forestry, horticulture and rural development at all administrative levels;

(*c*) Enhance popular participation in the management of local resources through appropriate legislation;

(*d*) Support non-governmental organizations and other private groups assisting local organizations and communities in the preparation of projects that would enhance participatory development of local people;

(*e*) Provide mechanisms to preserve threatened areas that could protect wildlife, conserve biological diversity or serve as national parks;

(*f*) Develop national policies that would provide incentives to farmers and local people to undertake conservation measures and to use environment-friendly technologies;

(*g*) Undertake income-generating activities in cottage and agro-processing industries, such as the cultivation and processing of medicinal and aromatic plants;

(*h*) Undertake the above activities, taking into account the need for full participation of women, including indigenous people and local communities, in development.

Data and Information

Governments at the appropriate level, with the support of the relevant international and regional organizations, should:

(*a*) Maintain and establish systematic observation and evaluation capacities at the national, state or provincial level to generate information for daily operations and to assess the environmental and socio-economic impacts of projects;

(*b*) Generate data on alternative livelihoods and diversified production systems at the village level on annual and tree crops, livestock, poultry, beekeeping, fisheries, village industries, markets, transport and income-earning opportunities, taking fully into account the role of women and integrating them into the planning and implementation process.

International and Regional Cooperation

Governments at the appropriate level, with the support of the relevant international and regional organizations, should:

(*a*) Strengthen the role of appropriate international research and training institutes such as the Consultative Group on International Agricultural Research Centers (CGIAR) and the International Board for Soil Research and Management (IBSRAM), as well as regional research centres, such as the Woodland Mountain Institutes and the International Center for Integrated Mountain Development, in undertaking applied research relevant to watershed development;

(*b*) Promote regional cooperation and exchange of data and information among countries sharing the same mountain ranges and river basins, particularly those affected by mountain disasters and floods;

(*c*) Maintain and establish partnerships with non-governmental organizations and other private groups working in watershed development.

Means of Implementation

Financial and Cost Evaluation

The Conference secretariat has estimated the average total annual cost (1993-2000) of implementing the activities of this programme to be about $13 billion, including about $1.9 billion from the international community on grant or concessional terms. These are indicative and order-ofmagnitude estimates only and have not been reviewed by Governments. Actual costs and financial terms, including any that are non-concessional, will depend upon, *inter alia*, the specific strategies and programmes Governments decide upon for implementation.

Financing for the promotion of alternative livelihoods in mountain ecosystems should be viewed as part of a country's anti-poverty or alternative livelihoods programme.

Scientific and Technical Means

Governments at the appropriate level, with the support of the relevant international and regional organizations, should:

(*a*) Consider undertaking pilot projects that combine environmental protection and development functions with particular emphasis on some of the traditional environmental management practices or systems that have a good impact on the environment;
(*b*) Generate technologies for specific watershed and farm conditions through a participatory approach involving local men and women, researchers and extension agents who will carry out experiments and trials on farm conditions;
(*c*) Promote technologies of vegetative conservation measures for erosion prevention, *in situ* moisture management, improved cropping technology, fodder production and agroforestry that are low-cost, simple and easily adopted by local people.

Human Resource Development

Governments at the appropriate level, with the support of the relevant international and regional organizations, should:

(*a*) Promote a multidisciplinary and cross-sectoral approach in training and the dissemination of knowledge to local people on a wide range of issues, such as household production systems, conservation and utilization of arable and non-arable land, treatment of drainage lines and recharging of groundwater, livestock management, fisheries, agroforestry and horticulture;
(*b*) Develop human resources by providing access to education, health, energy and infrastructure;
(*c*) Promote local awareness and preparedness for disaster prevention and mitigation, combined with the latest available technology for early warning and forecasting.

Capacity-building

Governments at the appropriate level, with the support of the relevant international and regional organizations, should develop and strengthen national centres for watershed management to encourage a comprehensive approach to the environmental, socio-economic, technological, legislative, financial and administrative aspects and provide support to policy makers, administrators, field staff and farmers for watershed development.

The private sector and local communities, in cooperation with national Governments, should promote local infrastructure development, including communication networks, mini- or micro-hydro development to support cottage industries, and access to markets.

PROMOTING SUSTAINABLE AGRICULTURE AND RURAL DEVELOPMENT

By the year 2025, 83 per cent of the expected global population of 8.5 billion will be living in developing countries. Yet the capacity of available resources and technologies to satisfy the demands of this growing population for food and other agricultural commodities remains uncertain. Agriculture has to meet this challenge, mainly by increasing production on land already in use and by avoiding further encroachment on land that is only marginally suitable for cultivation.

Major adjustments are needed in agricultural, environmental and macroeconomic policy, at both national and international levels, in developed as well as developing countries, to create the conditions for sustainable agriculture and rural development (SARD). The major objective of SARD is to increase food production in a sustainable way and enhance food security. This will involve education initiatives, utilization of economic incentives and the development of appropriate and new technologies, thus ensuring stable supplies of nutritionally adequate food, access to those supplies by vulnerable groups, and production for markets; employment and income generation to alleviate poverty; and natural resource management and environmental protection.

The priority must be on maintaining and improving the capacity of the higher potential agricultural lands to support an expanding population. However, conserving and rehabilitating the natural resources on lower potential lands in order to maintain sustainable man/land ratios is also necessary. The main tools of SARD are policy and agrarian reform, participation, income diversification, land conservation and improved management of inputs. The success of SARD will depend largely on the support and participation of rural people, national Governments, the private sector and international cooperation, including technical and scientific cooperation.

The following programme areas are included in this chapter:

(*a*) Agricultural policy review, planning and integrated programming in the light of the multifunctional aspect of agriculture, particularly with regard to food security and sustainable development;
(*b*) Ensuring people's participation and promoting human resource development for sustainable agriculture;
(*c*) Improving farm production and farming systems through diversification of farm and nonfarm employment and infrastructure development;
(*d*) Land-resource planning information and education for agriculture;
(*e*) Land conservation and rehabilitation;
(*f*) Water for sustainable food production and sustainable rural development;
(*g*) Conservation and sustainable utilization of plant genetic resources for food and sustainable agriculture;
(*h*) Conservation and sustainable utilization of animal genetic resources for sustainable agriculture;
(*i*) Integrated pest management and control in agriculture;
(*j*) Sustainable plant nutrition to increase food production;
(*k*) Rural energy transition to enhance productivity;
(*l*) Evaluation of the effects of ultraviolet radiation on plants and animals caused by the depletion of the stratospheric ozone layer.

Agricultural Policy Review, Planning and Integrated Programmes in the Light of the Multifunctional Aspect of Agriculture, Particularly with Regard to Food Security and Sustainable Development

Basis for Action

There is a need to integrate sustainable development considerations with agricultural policy analysis and planning in all countries, particularly in developing countries. Recommendations should contribute directly to development of realistic and operational medium- to long-term plans and programmes, and thus to concrete actions. Support to and monitoring of implementation should follow.

The absence of a coherent national policy framework for sustainable agriculture and rural development (SARD) is widespread and is not limited to the developing countries. In particular the economies in transition from planned to market-oriented systems need such a framework to incorporate environmental considerations into economic activities, including agriculture. All countries need to assess comprehensively the impacts of such policies on food and agriculture sector performance, food security, rural welfare and international trading relations as a means for identifying appropriate offsetting measures. The major thrust of food security in this case is to bring about a significant increase in agricultural production in a sustainable way and to achieve a substantial improvement in people's entitlement to adequate food and culturally appropriate food supplies.

Sound policy decisions pertaining to international trade and capital flows also necessitate action to overcome: (*a*) a lack of awareness of the environmental costs incurred by sectoral and macroeconomic policies and hence their threat to sustainability; (*b*) insufficient skills and experience in incorporating issues of sustainability into policies and programmes; and (*c*) inadequacy of tools of analysis and monitoring.

Objectives

The objectives of this programme area are:

(*a*) By 1995, to review and, where appropriate, establish a programme to integrate environmental and sustainable development with policy analysis for the food and agriculture sector and relevant macroeconomic policy analysis, formulation and implementation;

(*b*) To maintain and develop, as appropriate, operational multisectoral plans, programmes and policy measures, including programmes and measures to enhance sustainable food production and food security within the framework of sustainable development, not later than 1998;

(*c*) To maintain and enhance the ability of developing countries, particularly the least developed ones, to themselves manage policy, programming and planning activities, not later than 2005.

Activities

Management-related Activities

Governments at the appropriate level, with the support of the relevant international and regional organizations, should:

(*a*) Carry out national policy reviews related to food security, including adequate levels and stability of food supply and access to food by all households;

(*b*) Review national and regional agricultural policy in relation, *inter alia*, to foreign trade, price policy, exchange rate policies, agricultural subsidies and taxes, as well as organization for regional economic integration;

(*c*) Implement policies to influence land tenure and property rights positively with due recognition of the minimum size of land-holding required to maintain production and check further fragmentation;

(*d*) Consider demographic trends and population movements and identify critical areas for agricultural production;

(*e*) Formulate, introduce and monitor policies, laws and regulations and incentives leading to sustainable agricultural and rural development and improved food security and to the development and transfer of appropriate farm technologies, including, where appropriate, low-input sustainable agricultural (LISA) systems;

(*f*) Support national and regional early warning systems through food-security assistance schemes that monitor food supply and demand and factors affecting household access to food;

(*g*) Review policies with respect to improving harvesting, storage, processing, distribution and marketing of products at the local, national and regional levels;

(*h*) Formulate and implement integrated agricultural projects that include other natural resource activities, such as management of rangelands, forests, and wildlife, as appropriate;

(*i*) Promote social and economic research and policies that encourage sustainable agriculture development, particularly in fragile ecosystems and densely populated areas;

(*j*) Identify storage and distribution problems affecting food availability; support research, where necessary, to overcome these problems and cooperate with producers and distributors to implement improved practices and systems.

Data and Information

Governments at the appropriate level, with the support of the relevant international and regional organizations, should:

(*a*) Cooperate actively to expand and improve the information on early warning systems on food and agriculture at both regional and national levels;

(*b*) Examine and undertake surveys and research to establish baseline information on the status of natural resources relating to food and agricultural production and planning in order to assess the impacts of various uses on these resources, and develop methodologies and tools of analysis, such as environmental accounting.

International and Regional Cooperation and Coordination

United Nations agencies, such as FAO, the World Bank, IFAD and GATT, and regional organizations, bilateral donor agencies and other bodies should, within their respective mandates, assume a role in working with national Governments in the following activities:

(*a*) Implement integrated and sustainable agricultural development and food security strategies at the subregional level that use regional production and trade potentials, including organizations for regional economic integration, to promote food security;

(*b*) Encourage, in the context of achieving sustainable agricultural development and consistent with relevant internationally agreed principles on trade and environment, a more open and non-discriminatory trading system and the avoidance of unjustifiable trade barriers which together with other policies will facilitate the further integration of agricultural and environmental policies so as to make them mutually supportive;

(*c*) Strengthen and establish national, regional and international systems and networks to increase the understanding of the interaction between agriculture and the state of the environment, identify ecologically sound technologies and facilitate the exchange information on data sources, policies, and techniques and tools of analysis.

Means of Implementation

Financing and Cost Evaluation

The Conference secretariat has estimated the average total annual cost (1993-2000) on implementing the activities of this programme to be about $3 billion, including about $450 million from the international community on grant or concessional terms. These are indicative and order-ofmagnitude estimates only and have not been reviewed by Governments. Actual costs and financial terms, including any that are non-concessional, will depend upon, *inter alia*, the specific strategies and programmes Governments decide upon for implementation.

Scientific and Technological Means

Governments at the appropriate level and with the support of the relevant international and regional organizations should assist farming households and communities to apply technologies related to improved food production and security, including storage, monitoring of production and distribution.

Human Resource Development

Governments at the appropriate level, with the support of the relevant international and regional organizations, should:

(*a*) Involve and train local economists, planners and analysts to initiate national and international policy reviews and develop frameworks for sustainable agriculture;

(*b*) Establish legal measures to promote access of women to land and remove biases in their involvement in rural development.

Capacity-building

Governments at the appropriate level, with the support of the relevant international and regional organizations, should strengthen ministries for agriculture, natural resources and planning.

Ensuring People's Participation and Promoting Human Resource Development for Sustainable Agriculture

Basis for Action

This component bridges policy and integrated resource management. The greater the degree of community control over the resources on which it relies, the greater will be the incentive for economic and human resources development. At the same time, policy instruments to reconcile longrun and short-run requirements must be set by national Governments. The approaches focus on fostering self-reliance and cooperation, providing information and supporting user-based organizations. Emphasis should be on management practices, building agreements for changes in resource utilization, the rights and duties associated with use of land, water and forests, the functioning of markets, prices, and the access to information, capital and inputs. This would require training and capacity-building to assume greater responsibilities in sustainable development efforts.

Objectives

The objectives of this programme area are:

(*a*) To promote greater public awareness of the role of people's participation and people's organizations, especially women's groups, youth, indigenous people, local communities and small farmers, in sustainable agriculture and rural development;

(*b*) To ensure equitable access of rural people, particularly women, small farmers, landless and indigenous people, to land, water and forest resources and to technologies, financing, marketing, processing and distribution;

(*c*) To strengthen and develop the management and the internal capacities of rural people's organizations and extension services and to decentralize decision-making to the lowest community level.

Activities

Management-related Activities

Governments at the appropriate level, with the support of the relevant international and regional organizations, should:

(*a*) Develop and improve integrated agricultural extension services and facilities and rural organizations and undertake natural resource management and food security activities, taking into account the different needs of subsistence agriculture as well as market—oriented crops;

(*b*) Review and refocus existing measures to achieve wider access to land, water and forest resources and ensure equal rights of women and other disadvantaged groups, with particular emphasis on rural populations, indigenous people and local communities;

(*c*) Assign clear titles, rights and responsibilities for land and for individuals or communities to encourage investment in land resources;

(*d*) Develop guidelines for decentralization policies for rural development through reorganization and strengthening of rural institutions;

(*e*) Develop policies in extension, training, pricing, input distribution, credit and taxation to ensure necessary incentives and equitable access by the poor to production-support services;

(*f*) Provide support services and training, recognizing the variation in agricultural circumstances and practices by location; the optimal use of on-farm inputs and the minimal use of external inputs; optimal use of local natural resources and management of renewable energy sources; and the establishment of networks that deal with the exchange of information on alternative forms of agriculture.

Data and Information

Governments at the appropriate level, and with the support of the relevant international and regional organizations, should collect, analyse, and disseminate information on human resources, the role of Governments, local communities and non-governmental organizations in social innovation and strategies for rural development.

International and Regional Cooperation and Coordination

Appropriate international and regional agencies should:

(*a*) Reinforce their work with non-governmental organizations in collecting and disseminating information on people's participation and people's organizations, testing participatory development methods, training and education for human resource development and strengthening the management structures of rural organizations;

(*b*) Help develop information available through non-governmental organizations and promote an international ecological agricultural network to accelerate the development and implementation of ecological agriculture practices.

Means of Implementation

Financing and Cost Evaluation

The Conference secretariat has estimated the average total annual cost (1993-2000) of implementing the activities of this programme to be about $4.4 billion, including about $650 million from the international community on grant or concessional terms. These are indicative and order-ofmagnitude estimates only and have not been reviewed by Governments. Actual costs and financial terms, including any that are non-concessional, will depend upon, *inter alia*, the specific strategies and programmes Governments decide upon for implementation.

Scientific and Technological Means

Governments at the appropriate level, with the support of the relevant international and regional organizations, should:

(*a*) Encourage people's participation on farm technology development and transfer, incorporating indigenous ecological knowledge and practices;
(*b*) Launch applied research on participatory methodologies, management strategies and local organizations.

Human Resource Development

Governments at the appropriate level, with the support of the relevant international and regional organizations, should provide management and technical training to government administrators and members of resource-user groups in the principles, practice and benefits of people's participation in rural development.

Capacity-building

Governments at the appropriate level, with the support of the relevant international and regional organizations, should introduce management strategies and mechanisms, such as accounting and audit services for rural people's organizations and institutions for human resource development, and delegate administrative and financial responsibilities to local levels for decision-making, revenue raising and expenditure.

Improving Farm Production and Farming Systems through Diversification of Farm and Non-farm Employment and Infrastructure Development

Basis for Action

Agriculture needs to be intensified to meet future demands for commodities and to avoid further expansion onto marginal lands and encroachment on fragile ecosystems. Increased use of external inputs and development of specialized production and farming systems tend to increase vulnerability to environmental stresses and market fluctuations. There is, therefore, a need to intensify agriculture by diversifying the production systems for maximum efficiency in the utilization of local resources, while minimizing environmental and economic risks. Where intensification of farming systems is not possible, other on-farm and off-farm employment opportunities should be identified and developed, such as cottage industries, wildlife utilization, aquaculture and fisheries, non-farm activities, such as light village-based manufacturing, farm commodity processing, agribusiness, recreation and tourism, etc.

Objectives

The objectives of this programme area are:

(*a*) To improve farm productivity in a sustainable manner, as well as to increase diversification, efficiency, food security and rural incomes, while ensuring that risks to the ecosystem are minimized;

(*b*) To enhance the self-reliance of farmers in developing and improving rural infrastructure, and to facilitate the transfer of environmentally sound technologies for integrated production and farming systems, including indigenous technologies and the sustainable use of biological and ecological processes, including agroforestry, sustainable wildlife conservation and management, aquaculture, inland fisheries and animal husbandry;

(*c*) To create farm and non-farm employment opportunities, particularly among the poor and those living in marginal areas, taking into account the alternative livelihood proposal *inter alia* in dryland areas.

Activities

Management-related Activities

Governments at the appropriate level, with the support of the relevant international and regional organizations, should:

(*a*) Develop and disseminate to farming households integrated farm management technologies, such as crop rotation, organic manuring and other techniques involving reduced use of agricultural chemicals, multiple techniques for sources of nutrients and the efficient utilization of external inputs, while enhancing techniques for waste and byproduct utilization and prevention of pre- and post-harvest losses, taking particular note of the role of women;

(*b*) Create non-farm employment opportunities through private small-scale agro-processing units, rural service centres and related infrastructural improvements;

(*c*) Promote and improve rural financial networks that utilize investment capital resources raised locally;

(*d*) Provide the essential rural infrastructure for access to agricultural inputs and services, as well as to national and local markets, and reduce food losses;

(*e*) Initiate and maintain farm surveys, on-farm testing of appropriate technologies and dialogue with rural communities to identify constraints and bottlenecks and find solutions;

(*f*) Analyse and identify possibilities for economic integration of agricultural and forestry activities, as well as water and fisheries, and to take effective measures to encourage forest management and growing of trees by farmers (farm forestry) as an option for resource development.

Data and Information

Governments at the appropriate level, with the support of the relevant international and regional organizations, should:

(*a*) Analyse the effects of technical innovations and incentives on farm-household income and well-being;

(*b*) Initiate and maintain on-farm and off-farm programmes to collect and record indigenous knowledge.

International and Regional Cooperation and Coordination

International institutions, such as FAO and IFAD, international agricultural research centres, such as CGIAR, and regional centres should diagnose the world's major agro-ecosystems, their extension, ecological and socio-economic characteristics, their susceptibility to deterioration and their productive potential. This could form the basis for technology development and exchange and for regional research collaboration.

Means of Implementation

Financing and Cost Evaluation

The Conference secretariat has estimated the average total annual cost (1993-2000) of implementing the activities of this programme to be about $10 billion, including about $1.5 billion from the international community on grant or concessional terms. These are indicative and order-ofmagnitude estimates only and have not been reviewed by Governments. Actual costs and financial terms, including any that are non-concessional, will depend upon, *inter alia*, the specific strategies and programmes Governments decide upon for implementation.

Scientific and Technological Means

Governments at the appropriate level, with the support of the relevant international and regional organizations, should strengthen research on agricultural production systems in areas with different endowments and agro-ecological zones, including comparative analysis of the intensification, diversification and different levels of external and internal inputs.

Human Resource Development

Governments at the appropriate level, with the support of the relevant international and regional organizations, should:

(*a*) Promote educational and vocational training for farmers and rural communities through formal and non-formal education;
(*b*) Launch awareness and training programmes for entrepreneurs, managers, bankers and traders in rural servicing and small-scale agro-processing techniques.

Capacity-building

Governments at the appropriate level, with the support of the relevant international and regional organizations, should:

(*a*) Improve their organizational capacity to deal with issues related to off-farm activities and rural industry development;
(*b*) Expand credit facilities and rural infrastructure related to processing, transportation and marketing.

Land-resource Planning, Information and Education for Agriculture

Basis for Action

Inappropriate and uncontrolled land uses are a major cause of degradation and depletion of land resources. Present land use often disregards the actual potentials, carrying capacities and limitations of land resources, as well as their diversity in space. It is estimated that the world's population, now at 5.4 billion, will be 6.25 billion by the turn of the century. The need to increase food production to meet the expanding needs of the population will put enormous pressure on all natural resources, including land.

Poverty and malnutrition are already endemic in many regions. The destruction and degradation of agricultural and environmental resources is a major issue. Techniques for increasing production and conserving soil and water resources are already available but are not widely or systematically applied. A systematic approach is needed for identifying land uses and production systems that are sustainable in each land and climate zone, including the economic, social and institutional mechanisms necessary for their implementation.

Objectives

The objectives of this programme area are:

(*a*) To harmonize planning procedures, involve farmers in the planning process, collect landresource data, design and establish databases, define land areas of similar capability, identify resource problems and values that need to be taken into account to establish mechanisms to encourage efficient and environmentally sound use of resources;

(*b*) To establish agricultural planning bodies at national and local levels to decide priorities, channel resources and implement programmes.

Activities

Management-related Activities

Governments at the appropriate level, with the support of the relevant international and regional organizations, should:

(*a*) Establish and strengthen agricultural land-use and land-resource planning, management, education and information at national and local levels;

(*b*) Initiate and maintain district and village agricultural land-resource planning, management and conservation groups to assist in problem identification, development of technical and management solutions, and project implementation.

Data and Information

Governments at the appropriate level, with the support of the relevant international and regional organizations, should:

(*a*) Collect, continuously monitor, update and disseminate information, whenever possible, on the utilization of natural resources and living conditions, climate, water and soil factors, and on land use, distribution of vegetation cover and animal species, utilization of wild plants, production systems and yields, costs and prices, and social and cultural considerations that affect agricultural and adjacent land use;

(*b*) Establish programmes to provide information, promote discussion and encourage the formation of management groups.

International and Regional Cooperation and Coordination

The appropriate United Nations agencies and regional organizations should:

(*a*) Strengthen or establish international, regional and subregional technical working groups with specific terms of reference and budgets to promote the integrated use of land resources for agriculture, planning, data collection and diffusion of simulation models of production and information dissemination;
(*b*) Develop internationally acceptable methodologies for the establishment of databases, description of land uses and multiple goal optimization.

Means of Implementation

Financing and Cost Evaluation

The Conference secretariat has estimated the average total annual cost (1993-2000) of implementing the activities of this programme to be about $1.7 billion, including about $250 million from the international community on grant or concessional terms. These are indicative and order-ofmagnitude estimates only and have not been reviewed by Governments. Actual costs and financial terms, including any that are non-concessional, will depend upon, *inter alia*, the specific strategies and programmes Governments decide upon for implementation.

Scientific and Technological Means

Governments at the appropriate level, with the support of the relevant international and regional organizations, should:

(*a*) Develop databases and geographical information systems to store and display physical, social and economic information pertaining to agriculture, and the definition of ecological zones and devèlopment areas;
(*b*) Select combinations of land uses and production systems appropriate to land units through multiple goal

optimization procedures, and strengthen delivery systems and local community participation;

(*c*) Encourage integrated planning at the watershed and landscape level to reduce soil loss and protect surface and groundwater resources from chemical pollution.

Human Resource Development

Governments at the appropriate level, with the support of the relevant international and regional organizations, should:

(*a*) Train professionals and planning groups at national, district and village levels through formal and informal instructional courses, travel and interaction;

(*b*) Generate discussion at all levels on policy, development and environmental issues related to agricultural land use and management, through media programmes, conferences and seminars.

Capacity-building

Governments at the appropriate level, with the support of the relevant international and regional organizations, should:

(*a*) Establish land-resource mapping and planning units at national, district and village levels to act as focal points and links between institutions and disciplines, and between Governments and people;

(*b*) Establish or strengthen Governments and international institutions with responsibility for agricultural resource survey, management and development; rationalize and strengthen legal frameworks; and provide equipment and technical assistance.

Land Conservation and Rehabilitation

Basis for Action

Land degradation is the most important environmental problem affecting extensive areas of land in both developed and developing countries. The problem of soil erosion is particularly acute in developing countries, while problems of salinization, waterlogging, soil pollution and loss of soil fertility are increasing in all countries. Land degradation is serious because the productivity of huge areas of land is declining just when populations are increasing rapidly and the demand on the land is growing to produce more food, fibre and fuel. Efforts to control land degradation, particularly in developing countries, have had limited success to date. Well planned, long-term national and regional land conservation and rehabilitation programmes, with strong political support and adequate funding, are now needed. While land-use planning and land zoning, combined with better land management, should provide long-term solutions, it is urgent to arrest land degradation and launch conservation and rehabilitation programmes in the most critically affected and vulnerable areas.

Objectives

The objectives of this programme area are:

(*a*) By the year 2000, to review and initiate, as appropriate, national land-resource surveys, detailing the location, extent and severity of land degradation;

(*b*) To prepare and implement comprehensive policies and programmes leading to the reclamation of degraded lands and the conservation of areas at risk, as well as improve the general planning, management and utilization of land resources and preserve soil fertility for sustainable agricultural development.

Activities

Management-related Activities

Governments at the appropriate level, with the support of the relevant international and regional organizations, should:

(*a*) Develop and implement programmes to remove and resolve the physical, social and economic causes of land degradation, such as land tenure, appropriate trading systems and agricultural pricing structures, which lead to inappropriate land-use management;

(*b*) Provide incentives and, where appropriate and possible, resources for the participation of local communities in the planning, implementation and maintenance of their own conservation and reclamation programmes;

(*c*) Develop and implement programmes for the rehabilitation of land degraded by waterlogging and salinity;

(*d*) Develop and implement programmes for the progressive use of non-cultivated land with agricultural potential in a sustainable way.

Data and Information

Governments, at the appropriate level, with the support of the relevant international and regional organizations, should:

(*a*) Conduct periodic surveys to assess the extent and state of its land resources;

(*b*) Strengthen and establish national land-resource data banks, including identification of the location, extent and severity of existing land degradation, as well as areas at risk, and evaluate the progress of the conservation and rehabilitation programmes launched in this regard;

(*c*) Collect and record information on indigenous conservation and rehabilitation practices and farming systems as a basis for research and extension programmes.

International and Regional Cooperation and Coordination

The appropriate United Nations agencies, regional organizations and non-governmental organizations should:

(*a*) Develop priority conservation and rehabilitation programmes with advisory services to Governments and regional organizations;

(*b*) Establish regional and subregional networks for scientists and technicians to exchange experiences, develop joint programmes and spread successful technologies on land conservation and rehabilitation.

Means of Implementation

Financing and Cost Evaluation

The Conference secretariat has estimated the average total annual cost (1993-2000) of implementing the activities of this programme to be about $5 billion, including about $800 million from the international community on grant or concessional terms. These are indicative and order-ofmagnitude estimates only and have not been reviewed by Governments. Actual costs and financial terms, including any that are non-concessional, will depend upon, *inter alia*, the specific strategies and programmes Governments decide upon for implementation.

Scientific and Technological Means

Governments at the appropriate level, with the support of the relevant international and regional organizations, should help farming household communities to investigate and promote site-specific technologies and farming systems that conserve and rehabilitate land, while increasing agricultural production, including conservation tillage agroforestry, terracing and mixed cropping.

Human Resource Development

Governments at the appropriate level, with the support of the relevant international and regional organizations, should train field staff and land users in indigenous and modern techniques of conservation and rehabilitation and should establish training facilities for extension staff and land users.

Capacity-building

Governments at the appropriate level, with the support of the relevant international and regional organizations, should:

(*a*) Develop and strengthen national research institutional capacity to identify and implement effective conservation and rehabilitation practices that are appropriate to the existing socio-economic physical conditions of the land users;

(*b*) Coordinate all land conservation and rehabilitation policies, strategies and programmes with related ongoing programmes, such as national environment action plans, the Tropical Forestry Action Plan and national development programmes.

Water for Sustainable Food Production and Sustainable Rural Development

This programme area is included (Protection of the quality and supply of freshwater resources).

Conservation and Sustainable Utilization of Plant Genetic Resources for Food and Sustainable Agriculture

Basis for Action

Plant genetic resources for agriculture (PGRFA) are an essential resource to meet future needs for food. Threats to the security of these resources are growing, and efforts to conserve, develop and use genetic diversity are underfunded and understaffed. Many existing gene banks provide inadequate security and, in some instances, the loss of plant genetic diversity in gene banks is as great as it is in the field.

The primary objective is to safeguard the world's genetic resources while preserving them to use sustainably. This includes the development of measures to facilitate the conservation and use of plant genetic resources, networks of *in situ* conservation areas and use of tools such as *ex situ* collections and germ plasma banks. Special emphasis could be placed on the building of endogenous capacity for characterization, evaluation and utilization of PGRFA, particularly for the minor crops and other underutilized or non-utilized species of food and agriculture, including tree species for agro-forestry. Subsequent action could be aimed at consolidation and efficient management of networks of *in situ* conservation areas and use of tools such as *ex situ* collections and germ plasma banks.

Major gaps and weaknesses exist in the capacity of existing national and international mechanisms to assess, study, monitor and use plant genetic resources to increase food production. Existing institutional capacity, structures and programmes are generally inadequate and largely underfunded. There is genetic erosion of invaluable crop species. Existing diversity in crop species is not used to the extent possible for increased food production in a sustainable way.

Objectives

The objectives of this programme area are:

(*a*) To complete the first regeneration and safe duplication of existing *ex situ* collections on a world-wide basis as soon as possible;
(*b*) To collect and study plants useful for increasing food production through joint activities, including training, within the framework of networks of collaborating institutions;
(*c*) Not later than the year 2000, to adopt policies and strengthen or establish programmes for *in situ* on-farm and *ex situ* conservation and sustainable use of plant genetic resources for food and agriculture, integrated into strategies and programmes for sustainable agriculture;
(*d*) To take appropriate measures for the fair and equitable sharing of benefits and results of research and development in plant breeding between the sources and users of plant genetic resources.

Activities

Management-related Activities

Governments at the appropriate level, with the support of the relevant international and regional organizations, should:

(*a*) Develop and strengthen institutional capacity, structures and programmes for conservation and use of PGRFA;
(*b*) Strengthen and establish research in the public domain on PGRFA evaluation and utilization, with the objectives of

sustainable agriculture and rural development in view;

(*c*) Develop multiplication/propagation, exchange and dissemination facilities for PGRFAs (seeds and planting materials), particularly in developing countries and monitor, control and evaluate plant introductions;

(*d*) Prepare plans or programmes of priority action on conservation and sustainable use of PGRFA, based, as appropriate, on country studies on PGRFA;

(*e*) Promote crop diversification in agricultural systems where appropriate, including new plants with potential value as food crops;

(*f*) Promote utilization as well as research on poorly known, but potentially useful, plants and crops, where appropriate;

(*g*) Strengthen national capabilities for utilization of PGRFA, plant breeding and seed production capabilities, both by specialized institutions and farming communities.

Data and Information

Governments at the appropriate level, with the support of the relevant international and regional organizations, should:

(*a*) Develop strategies for networks of *in situ* conservation areas and use of tools such as onfarm *ex situ* collections, germplasm banks and related technologies;

(*b*) Establish *ex situ* base collection networks;

(*c*) Review periodically and report on the situation on PGRFA, using existing systems and procedures;

(*d*) Characterize and evaluate PGRFA material collected, disseminate information to facilitate the use of PGRFA collections and assess genetic variation in collections.

International and Regional Cooperation and Coordination

The appropriate United Nations agencies and regional organizations should:

(*a*) Strengthen the Global System on the Conservation and Sustainable Use of PGRFA by, *inter alia*, accelerating the development of the Global Information and Early Warning

System to facilitate the exchange of information; developing ways to promote the transfer of environmentally sound technologies, in particular to developing countries; and taking further steps to realize farmers' rights;

(*b*) Develop subregional, regional and global networks of PGRFA *in situ* in protected areas;

(*c*) Prepare periodic state of the world reports on PGRFA;

(*d*) Prepare a rolling global cooperative plan of action on PGRFA;

(*e*) Promote, for 1994, the Fourth International Technical Conference on the Conservation and Sustainable Use of PGRFA, which is to adopt the first state of the world report and the first global plan of action on the conservation and sustainable use of PGRFA;

(*f*) Adjust the Global System for the Conservation and Sustainable Use of PGRFA in line with the outcome of the negotiations of a convention on biological diversity.

Means of Implementation

Financing and Cost Evaluation

The Conference secretariat has estimated the average total annual cost (1993-2000) of implementing the activities of this programme to be about $600 million, including about $300 million from the international community on grant or concessional terms. These are indicative and order-of-magnitude estimates only and have not been reviewed by Governments. Actual costs and financial terms, including any that are non-concessional, will depend upon, *inter alia*, the specific strategies and programmes Governments decide upon for implementation.

Scientific and Technological Means

Governments, at the appropriate level, with the support of the relevant international and regional organizations, should:

(*a*) Develop basic science research in such areas as plant taxonomy and phytogeography, utilizing recent developments, such as computer sciences, molecular genetics and in vitro cryopreservation;

(*b*) Develop major collaborative projects between research programmes in developed and developing countries, particularly for the enhancement of poorly known or neglected crops;

(*c*) Promote cost-effective technologies for keeping duplicate sets of *ex situ* collections (which can also be used by local communities);

(*d*) Develop further conservation sciences in relation to *in situ* conservation and technical means to link it with *ex situ* conservation efforts.

Human Resource Development

Governments at the appropriate level and with the support of the relevant international and regional organizations should:

(*a*) Promote training programmes at both undergraduate and post-graduate levels in conservation sciences for running PGRFA facilities and for the design and implementation of national programmes in PGRFA;

(*b*) Raise the awareness of agricultural extension services in order to link PGRFA activities with user communities;

(*c*) Develop training materials to promote conservation and utilization of PGRFA at the local level.

Capacity-building

Governments at the appropriate level, with the support of the relevant international and regional organizations, should establish national policies to provide legal status for and strengthen legal aspects of PGRFA, including long-term financial commitments for germplasm collections and implementation of activities in PGRFA.

Conservation and Sustainable Utilization of Animal Genetic Resources for Sustainable Agriculture

Basis for Action

The need for increased quantity and quality of animal products and for draught animals calls for conservation of the existing diversity of animal breeds to meet future requirements, including those for use in biotechnology. Some local animal breeds, in addition to their socio-cultural value, have unique attributes for adaptation, disease resistance and specific uses and should be preserved. These local breeds are threatened by extinction as a result of the introduction of exotic breeds and of changes in livestock production systems.

Objectives

The objectives of this programme area are:

(*a*) To enumerate and describe all breeds of livestock used in animal agriculture in as broad a way as possible and begin a 10-year programme of action;
(*b*) To establish and implement action programmes to identify breeds at risk, together with the nature of the risk and appropriate preservation measures;
(*c*) To establish and implement development programmes for indigenous breeds in order to guarantee their survival, avoiding the risk of their being replaced by breed substitution or cross-breeding programmes.

Activities

Management-related Activities

Governments at the appropriate level, with the support of the relevant international and regional organizations, should:

(*a*) Draw up breed preservation plans, for endangered populations, including semen/embryo collection and storage, farm-based conservation of indigenous stock or *in situ* preservation;
(*b*) Plan and initiate breed development strategies;
(*c*) Select indigenous populations on the basis of regional importance and genetic uniqueness, for a 10-year programme, followed by selection of an additional cohort of indigenous breeds for development.

Data and Information

Governments at the appropriate level, with the support of the relevant international and regional organizations, should prepare and complete national inventories of available animal genetic resources. Cryogenic storage could be given priority over characterization and evaluation. Training of nationals in conservation and assessment techniques would be given special attention.

International and Regional Cooperation and Coordination

The appropriate United Nations and other international and regional agencies should:

(*a*) Promote the establishment of regional gene banks to the extent that they are justified, based on principles of technical cooperation among developing countries;

(*b*) Process, store and analyse animal genetic data at the global level, including the establishment of a world watch list and an early warning system for endangered breeds; global assessment of scientific and intergovernmental guidance of the programme and review of regional and national activities; development of methodologies, norms and standards (including international agreements); monitoring of their implementation; and related technical and financial assistance;

(*c*) Prepare and publish a comprehensive database of animal genetic resources, describing each breed, its derivation, its relationship with other breeds, effective population size and a concise set of biological and production characteristics;

(*d*) Prepare and publish a world watch list on farm animal species at risk to enable national Governments to take action to preserve endangered breeds and to seek technical assistance, where necessary.

Means of Implementation

Financing and Cost Evaluation

The Conference secretariat has estimated the average total annual cost (1993-2000) of implementing the activities of this programme

to be about $200 million, including about $100 million from the international community on grant or concessional terms. These are indicative and order-of-magnitude estimates only and have not been reviewed by Governments. Actual costs and financial terms, including any that are non-concessional, will depend upon, *inter alia*, the specific strategies and programmes Governments decide upon for implementation.

Scientific and Technological Means

Governments at the appropriate level, with the support of the relevant international and regional organizations, should:

(*a*) Use computer-based data banks and questionnaires to prepare a global inventory/world watch list;
(*b*) Using cryogenic storage of germplasm, preserve breeds at serious risk and other material from which genes can be reconstructed.

Human Resource Development

Governments at the appropriate level, with the support of the relevant international and regional organizations, should:

(*a*) Sponsor training courses for nationals to obtain the necessary expertise for data collection and handling and for the sampling of genetic material;
(*b*) Enable scientists and managers to establish an information base for indigenous livestock breeds and promote programmes to develop and conserve essential livestock genetic material.

Capacity-building

Governments at the appropriate level, with the support of the relevant international and regional organizations, should:

(*a*) Establish in-country facilities for artificial insemination centres and *in situ* breeding farms;

(*b*) Promote in-country programmes and related physical infrastructure for animal livestock conservation and breed development, as well as for strengthening national capacities to take preventive action when breeds are endangered.

Integrated Pest Management and Control in Agriculture

Basis for Action

World food demand projections indicate an increase of 50 per cent by the year 2000 which will more than double again by 2050. Conservative estimates put pre-harvest and post-harvest losses caused by pests between 25 and 50 per cent. Pests affecting animal health also cause heavy losses and in many areas prevent livestock development. Chemical control of agricultural pests has dominated the scene, but its overuse has adverse effects on farm budgets, human health and the environment, as well as on international trade. New pest problems continue to develop. Integrated pest management, which combines biological control, host plant resistance and appropriate farming practices and minimizes the use of pesticides, is the best option for the future, as it guarantees yields, reduces costs, is environmentally friendly and contributes to the sustainability of agriculture. Integrated pest management should go hand in hand with appropriate pesticide management to allow for pesticide regulation and control, including trade, and for the safe handling and disposal of pesticides, particularly those that are toxic and persistent.

Objectives

The objectives of this programme area are:

(*a*) Not later than the year 2000, to improve and implement plant protection and animal health services, including mechanisms to control the distribution and use of pesticides, and to implement the International Code of Conduct on the Distribution and Use of Pesticides;

(*b*) To improve and implement programmes to put integrated pest-management practices within the reach of farmers through farmer networks, extension services and research institutions;

(*c*) Not later than the year 1998, to establish operational and interactive networks among farmers, researchers and extension services to promote and develop integrated pest-management.

Activities

Management-related Activities

Governments at the appropriate level, with the support of the relevant international and regional organizations, should:

(*a*) Review and reform national policies and the mechanisms that would ensure the safe and appropriate use of pesticides—for example, pesticide pricing, pest control brigades, price structure of inputs and outputs and integrated pest-management policies and action plans;

(*b*) Develop and adopt efficient management systems to control and monitor the incidence of pests and disease in agriculture and the distribution and use of pesticides at the country level;

(*c*) Encourage research and development into pesticides that are target-specific and readily degrade into harmless constituent parts after use;

(*d*) Ensure that pesticide labels provide farmers with understandable information about safe handling, application and disposal.

Data and Information

Governments at the appropriate level, with the support of the relevant international and regional organizations, should:

(*a*) Consolidate and harmonize existing information and programmes on the use of pesticides that have been banned or severely restricted in different countries;

(*b*) Consolidate, document and disseminate information on biological control agents and organic pesticides, as well as on traditional and other relevant knowledge and skills regarding alternative non-chemical ways of controlling pests;

(*c*) Undertake national surveys to establish baseline information on the use of pesticides in each country and the side-effects on human health and environment, and also undertake appropriate education.

International and Regional Cooperation and Coordination

Appropriate United Nations agencies and regional organizations should:

(*a*) Establish a system for collecting, analysing and disseminating data on the quantity and quality of pesticides used every year and their impact on human health and the environment;

(*b*) Strengthen regional interdisciplinary projects and establish integrated pest management (IPM) networks to demonstrate the social, economic and environmental benefits of IPM for food and cash crops in agriculture;

(*c*) Develop proper IPM, comprising the selection of the variety of biological, physical and cultural controls, as well as chemical controls, taking into account specific regional conditions.

Means of Implementation

Financing and Cost Evaluation

The Conference secretariat has estimated the average total annual cost (1993-2000) of implementing the activities of this programme to be about $1.9 billion, including about $285 million from the international community on grant or concessional terms. These are indicative and order-ofmagnitude estimates only and have not been reviewed by Governments. Actual costs and financial terms, including any that are non-concessional, will depend upon, *inter alia*, the specific strategies and programmes Governments decide upon for implementation.

Scientific and Technological Means

Governments at the appropriate level, with the support of the relevant international and regional organizations, should launch on-farm research in the development of non-chemical alternative pest management technologies.

Human Resource Development

Governments at the appropriate level, with the support of the relevant international and regional organizations, should:

(*a*) Prepare and conduct training programmes on approaches and techniques for integrated pest management and control of pesticide use, to inform policy makers, researchers, non-governmental organizations and farmers;

(*b*) Train extension agents and involve farmers and women's groups in crop health and alternative non-chemical ways of controlling pests in agriculture.

Capacity-building

Governments at the appropriate level, with the support of the relevant international and regional organizations, should strengthen national public administrations and regulatory bodies in the control of pesticides and the transfer of technology for integrated pest management.

Sustainable Plant Nutrition to Increase Food Production

Basis for Action

Plant nutrient depletion is a serious problem resulting in loss of soil fertility, particularly in developing countries. To maintain soil productivity, the FAO sustainable plant nutrition programmes could be helpful. In sub-Saharan Africa, nutrient output from all sources currently exceeds inputs by a factor of three or four, the net loss being estimated at some 10 million metric tons per year. As a result, more marginal lands and fragile natural ecosystems are put under agricultural use, thus creating further land degradation and other environmental problems. The integrated plant nutrition approach aims at ensuring a sustainable supply of plant nutrients to increase future yields without harming the environment and soil productivity.

In many developing countries, population growth rates exceed 3 per cent a year, and national agricultural production has fallen behind food demand. In these countries the goal should be to increase agricultural production by at least 4 per cent a year, without destroying the soil fertility. This will require increasing agricultural production in high-potential areas through efficiency in the use of inputs. Trained labour, energy supply, adapted tools and technologies, plant nutrients and soil enrichment will all be essential.

Objectives

The objectives of this programme area are:

(*a*) Not later than the year 2000, to develop and maintain in all countries the integrated plant nutrition approach, and to optimize availability of fertilizer and other plant nutrient sources;

(*b*) Not later than the year 2000, to establish and maintain institutional and human infrastructure to enhance effective decision-making on soil productivity;

(*c*) To develop and make available national and international know-how to farmers, extension agents, planners and policy makers on environmentally sound new and existing technologies and soil-fertility management strategies for application in promoting sustainable agriculture.

Activities

Management-related Activities

Governments at the appropriate level, with the support of the relevant international and regional organizations, should:

(*a*) Formulate and apply strategies that will enhance soil fertility maintenance to meet sustainable agricultural production and adjust the relevant agricultural policy instruments accordingly;

(*b*) Integrate organic and inorganic sources of plant nutrients in a system to sustain soil fertility and determine mineral fertilizer needs;

(*c*) Determine plant nutrient requirements and supply strategies and optimize the use of both organic and inorganic sources, as appropriate, to increase farming efficiency and production;

(*d*) Develop and encourage processes for the recycling of organic and inorganic waste into the soil structure, without harming the environment, plant growth and human health.

Data and Information

Governments at the appropriate level, with the support of the relevant international and regional organizations, should:

(*a*) Assess "national accounts" for plant nutrients, including supplies (inputs) and losses (outputs) and prepare balance sheets and projections by cropping systems;

(*b*) Review technical and economic potentials of plant nutrient sources, including national deposits, improved organic supplies, recycling, wastes, topsoil produced from discarded organic matter and biological nitrogen fixation.

International and Regional Cooperation and Coordination

The appropriate United Nations agencies, such as FAO, the international agricultural research institutes, and non-governmental organizations should collaborate in carrying out information and publicity campaigns about the integrated plant nutrients approach, efficiency of soil productivity and their relationship to the environment.

Means of Implementation

Financing and Cost Evaluation

The Conference secretariat has estimated the average total annual cost (1993-2000) of implementing the activities of this programme to be about $3.2 billion, including about $475 million from the international community on grant or concessional terms. These are indicative and order-of magnitude estimates only and have not been reviewed by Governments. Actual costs and financial terms, including any that are non-concessional, will depend upon, *inter alia*, the specific strategies and programmes Governments decide upon for implementation.

Scientific and Technological Means

Governments at the appropriate level, with the support of the relevant international and regional organizations, should:

(*a*) Develop site-specific technologies at benchmark sites and farmers' fields that fit prevailing socio-economic and ecological conditions through research that involves the full collaboration of local populations;

(*b*) Reinforce interdisciplinary international research and transfer of technology in cropping and farming systems research, improved *in situ* biomass production techniques, organic residue management and agroforestry technologies.

Human Resource Development

Governments at the appropriate level, with the support of the relevant international and regional organizations, should:

(*a*) Train extension officers and researchers in plant nutrient management, cropping systems and farming systems, and in economic evaluation of plant nutrient impact;

(*b*) Train farmers and women's groups in plant nutrition management, with special emphasis on topsoil conservation and production.

Capacity-building

Governments at the appropriate level, with the support of the relevant international and regional organizations, should:

(*a*) Develop suitable institutional mechanisms for policy formulation to monitor and guide the implementation of integrated plant nutrition programmes through an interactive process involving farmers, research, extension services and other sectors of society;

(*b*) Where appropriate, strengthen existing advisory services and train staff, develop and test new technologies and facilitate the adoption of practices to upgrade and maintain full productivity of the land.

Rural Energy Transition to Enhance Productivity

Basis for Action

Energy supplies in many countries are not commensurate with their development needs and are highly priced and unstable. In rural areas of the developing countries, the chief sources of energy are fuelwood, crop residues and manure, together with animal and human energy. More intensive energy inputs are required for increased productivity of human labour and for income-generation. To this end, rural energy policies and technologies should promote a mix of cost-effective fossil and renewable energy sources that is itself sustainable and ensures sustainable agricultural development. Rural areas provide energy supplies in the form of wood. The full potential of agriculture and agroforestry, as well as common property resources, as sources of renewable energy, is far from being realized. The attainment of sustainable rural development is intimately linked with energy demand and supply patterns.

Objectives

The objectives of this programme area are:

(*a*) Not later than the year 2000, to initiate and encourage a process of environmentally sound energy transition in rural communities, from unsustainable energy sources, to structured and diversified energy sources by making available alternative new and renewable sources of energy;

(*b*) To increase the energy inputs available for rural household and agro-industrial needs through planning and appropriate technology transfer and development;

(*c*) To implement self-reliant rural programmes favouring sustainable development of renewable energy sources and improved energy efficiency.

Activities

Management-related Activities

Governments at the appropriate level, with the support of the relevant international and regional organizations, should:

(*a*) Promote pilot plans and projects consisting of electrical, mechanical and thermal power (gasifiers, biomass, solar driers, wind-pumps and combustion systems) that are appropriate and likely to be adequately maintained;
(*b*) Initiate and promote rural energy programmes supported by technical training, banking and related infrastructure;
(*c*) Intensify research and the development, diversification and conservation of energy, taking into account the need for efficient use and environmentally sound technology.

Data and Information

Governments at the appropriate level, with the support of the relevant international and regional organizations, should:

(*a*) Collect and disseminate data on rural energy supply and demand patterns related to energy needs for households, agriculture and agro-industry;
(*b*) Analyse sectoral energy and production data in order to identify rural energy requirements.

International and Regional Cooperation and Coordination

The appropriate United Nations agencies and regional organizations should, drawing on the experience and available information of non-governmental organizations in this field, exchange country and regional experience on rural energy planning methodologies in order to promote efficient planning and select cost-effective technologies.

Means of Implementation

Financing and Cost Evaluation

The Conference secretariat has estimated the average total annual cost (1993-2000) of implementing the activities of this programme to be about $1.8 billion per year, including about $265 million from the international community on grant or concessional terms. These are indicative and order-of-magnitude estimates only and have not been reviewed by Governments. Actual costs and financial terms, including any that are non-concessional, will depend upon, *inter alia*, the specific strategies and programmes Governments decide upon for implementation.

Scientific and Technological Means

Governments at the appropriate level, with the support of the relevant international and regional organizations, should:

(*a*) Intensify public and private sector research in developing and industrialized countries on renewable sources of energy for agriculture;
(*b*) Undertake research and transfer of energy technologies in biomass and solar energy to agricultural production and post-harvest activities.

Human Resource Development

Governments at the appropriate level, with the support of the relevant international and regional organizations, should enhance public awareness of rural energy problems, stressing the economic and environmental advantages of renewable energy sources.

Capacity-building

Governments at the appropriate level, with the support of the relevant international and regional organizations, should:

(*a*) Establish national institutional mechanisms for rural energy planning and management that would improve efficiency in agricultural productivity and reach the village and household level;

(*b*) Strengthen extension services and local organizations to implement plans and programmes for new and renewable sources of energy at the village level.

Evaluation of the Effects of Ultraviolet Radiation on Plants and Animals Caused by the Depletion of the Stratospheric Ozone Layer

Basis for Action

The increase of ultraviolet radiation as a consequence of the depletion of the stratospheric ozone layer is a phenomenon that has been recorded in different regions of the world, particularly in the southern hemisphere. Consequently, it is important to evaluate its effects on plant and animal life, as well as on sustainable agricultural development.

Objective

The objective of this programme area is to undertake research to determine the effects of increased ultraviolet radiation resulting from stratospheric ozone layer depletion on the Earth's surface, and on plant and animal life in affected regions, as well as its impact on agriculture, and to develop, as appropriate, strategies aimed at mitigating its adverse effects.

Activities

Management-related Activities

In affected regions, Governments at the appropriate level, with the support of the relevant international and regional organizations, should take the necessary measures, through institutional cooperation, to facilitate the implementation of research and evaluation regarding the effects of enhanced ultraviolet radiation on plant and animal life, as well as on agricultural activities, and consider taking appropriate remedial measures.

CONSERVATION OF BIOLOGICAL DIVERSITY

The objectives and activities in this chapter of Agenda 21 are intended to improve the conservation of biological diversity and the sustainable use of biological resources, as well as to support the Convention on Biological Diversity.

Our planet's essential goods and services depend on the variety and variability of genes, species, populations and ecosystems. Biological resources feed and clothe us and provide housing, medicines and spiritual nourishment. The natural ecosystems of forests, savannahs, pastures and rangelands, deserts, tundras, rivers, lakes and seas contain most of the Earth's biodiversity. Farmers' fields and gardens are also of great importance as repositories, while gene banks, botanical gardens, zoos and other germplasm repositories make a small but significant contribution. The current decline in biodiversity is largely the result of human activity and represents a serious threat to human development.

Conservation of Biological Diversity

Basis for Action

Despite mounting efforts over the past 20 years, the loss of the world's biological diversity, mainly from habitat destruction, over-harvesting, pollution and the inappropriate introduction of foreign plants and animals, has continued. Biological resources constitute a capital asset with great potential for yielding sustainable benefits. Urgent and decisive action is needed to conserve and maintain genes, species and ecosystems, with a view to the sustainable management and use of biological resources. Capacities for the assessment, study and systematic observation and evaluation of biodiversity need to be reinforced at national and international levels. Effective national action and international cooperation is required for the *in situ* protection of ecosystems, for the *ex situ* conservation of biological and genetic resources and for the enhancement of ecosystem functions. The participation and support of local communities are elements essential to the success of such an approach. Recent advances in biotechnology have pointed up the likely potential for agriculture, health and welfare and for the environmental purposes of the genetic material contained in plants, animals and micro-organisms. At the same time, it is particularly important in this context to stress that States have the sovereign right to exploit their own biological resources pursuant to their environmental policies, as well as the responsibility to conserve their biodiversity and use their biological resources sustainably, and to ensure that activities within their jurisdiction or control do not cause damage to the biological diversity of other States or of areas beyond the limits of national jurisdiction.

Objectives

Governments at the appropriate level, with the cooperation of the relevant United Nations bodies and regional, intergovernmental and non-governmental organizations, the private sector and financial institutions, and taking into consideration indigenous people and their communities, as well as social and economic factors, should:

(*a*) Press for the early entry into force of the Convention on Biological Diversity, with the widest possible participation;
(*b*) Develop national strategies for the conservation of biological diversity and the sustainable use of biological resources;
(*c*) Integrate strategies for the conservation of biological diversity and the sustainable use of biological resources into national development strategies and/or plans;
(*d*) Take appropriate measures for the fair and equitable sharing of benefits derived from research and development and use of biological and genetic resources, including biotechnology, between the sources of those resources and those who use them;
(*e*) Carry out country studies, as appropriate, on the conservation of biological diversity and the sustainable use of biological resources, including analyses of relevant costs and benefits, with particular reference to socio-economic aspects;
(*f*) Produce regularly updated world reports on biodiversity based upon national assessments;
(*g*) Recognize and foster the traditional methods and the knowledge of indigenous people and their communities,

emphasizing the particular role of women, relevant to the conservation of biological diversity and the sustainable use of biological resources, and ensure the opportunity for the participation of those groups in the economic and commercial benefits derived from the use of such traditional methods and knowledge;

(*h*) Implement mechanisms for the improvement, generation, development and sustainable use of biotechnology and its safe transfer, particularly to developing countries, taking account the potential contribution of biotechnology to the conservation of biological diversity and the sustainable use of biological resources;

(*i*) Promote broader international and regional cooperation in furthering scientific and economic understanding of the importance of biodiversity and its functions in ecosystems;

(*j*) Develop measures and arrangements to implement the rights of countries of origin of genetic resources or countries providing genetic resources, as defined in the Convention on Biological Diversity, particularly developing countries, to benefit from the biotechnological development and the commercial utilization of products derived from such resources.

Activities

Management-related Activities

Governments at the appropriate levels, consistent with national policies and practices, with the cooperation of the relevant United Nations bodies and, as appropriate, intergovernmental organizations and, with the support of indigenous people and their communities, non-governmental organizations and other groups, including the business and scientific communities, and consistent with the requirements of international law, should, as appropriate:

(*a*) Develop new or strengthen existing strategies, plans or programmes of action for the conservation of biological diversity and the sustainable use of biological resources, taking account of education and training needs;

(*b*) Integrate strategies for the conservation of biological diversity and the sustainable use of biological and genetic resources into relevant sectoral or cross-sectoral plans, programmes and policies, with particular reference to the special importance of terrestrial and aquatic biological and genetic resources for food and agriculture;

(*c*) Undertake country studies or use other methods to identify components of biological diversity important for its conservation and for the sustainable use of biological resources, ascribe values to biological and genetic resources, identify processes and activities with significant impacts upon biological diversity, evaluate the potential economic implications of the conservation of biological diversity and the sustainable use of biological and genetic resources, and suggest priority action;

(*d*) Take effective economic, social and other appropriate incentive measures to encourage the conservation of biological diversity and the sustainable use of biological resources, including the promotion of sustainable production systems, such as traditional methods of agriculture, agroforestry, forestry, range and wildlife management, which use, maintain or increase biodiversity;

(*e*) Subject to national legislation, take action to respect, record, protect and promote the wider application of the knowledge, innovations and practices of indigenous and local communities embodying traditional lifestyles for the conservation of biological diversity and the sustainable use of biological resources, with a view to the fair and equitable sharing of the benefits arising, and promote mechanisms to involve those communities, including women, in the conservation and management of ecosystems;

(*f*) Undertake long-term research into the importance of biodiversity for the functioning of ecosystems and the role of ecosystems in producing goods, environmental services and other values supporting sustainable development, with particular reference to the biology and reproductive capacities of key terrestrial and aquatic species, including native, cultivated and cultured species; new observation and inventory techniques; ecological conditions necessary for biodiversity conservation and continued evolution; and social behaviour and nutrition habits dependent on natural ecosystems, where women play key roles. The work should be undertaken with the widest possible participation, especially of indigenous people and their communities, including women;

(*g*) Take action where necessary for the conservation of biological diversity through the *in situ* conservation of ecosystems and natural habitats, as well as primitive cultivars and their wild relatives, and the maintenance and recovery of viable populations of species in their natural surroundings, and implement *ex situ* measures, preferably in the source country. *In situ* measures should include the reinforcement of terrestrial, marine and aquatic protected area systems and embrace, *inter alia*, vulnerable freshwater and other wetlands and coastal ecosystems, such as estuaries, coral reefs and mangroves;

(*h*) Promote the rehabilitation and restoration of damaged ecosystems and the recovery of threatened and endangered species;

(*i*) Develop policies to encourage the conservation of biodiversity and the sustainable use of biological and genetic resources on private lands;

(*j*) Promote environmentally sound and sustainable development in areas adjacent to protected areas with a view to furthering protection of these areas;

(*k*) Introduce appropriate environmental impact assessment procedures for proposed projects likely to have significant impacts upon biological diversity, providing for suitable information to be made widely available and for public participation, where appropriate, and encourage the assessment of the impacts of relevant policies and programmes on biological diversity;

(*l*) Promote, where appropriate, the establishment and strengthening of national inventory, regulation or management and control systems related to biological resources, at the appropriate level;

(*m*) Take measures to encourage a greater understanding and appreciation of the value of biological diversity, as manifested both in its component parts and in the ecosystem services provided.

Data and Information

Governments at the appropriate level, consistent with national policies and practices, with the cooperation of the relevant United Nations bodies and, as appropriate, intergovernmental organizations, and with the support of indigenous people and their communities, non-governmental organizations and other groups, including the business and scientific communities, and consistent with the requirements of international law, should, as appropriate:

(*a*) Regularly collate, evaluate and exchange information on the conservation of biological diversity and the sustainable use of biological resources;
(*b*) Develop methodologies with a view to undertaking systematic sampling and evaluation on a national basis of the components of biological diversity identified by means of country studies;
(*c*) Initiate or further develop methodologies and begin or continue work on surveys at the appropriate level on the status of ecosystems and establish baseline information on biological and genetic resources, including those in terrestrial, aquatic, coastal and marine ecosystems, as well as inventories undertaken with the participation of local and indigenous people and their communities;
(*d*) Identify and evaluate the potential economic and social implications and benefits of the conservation and sustainable use of terrestrial and aquatic species in each country, building upon the results of country studies;
(*e*) Undertake the updating, analysis and interpretation of data derived from the identification, sampling and evaluation activities described above;
(*f*) Collect, assess and make available relevant and reliable information in a timely manner and in a form suitable for decision-making at all levels, with the full support and participation of local and indigenous people and their communities.

International and Regional Cooperation and Coordination

Governments at the appropriate level, with the cooperation of the relevant United Nations bodies and, as appropriate, intergovernmental organizations, and, with the support of indigenous people and their communities, non-governmental organizations and other groups, including the business and scientific communities, and consistent with the requirements of international law, should, as appropriate:

(*a*) Consider the establishment or strengthening of national or international capabilities and networks for the exchange of data and information of relevance to the conservation of biological diversity and the sustainable use of biological and genetic resources;
(*b*) Produce regularly updated world reports on biodiversity based upon national assessments in all countries;
(*c*) Promote technical and scientific cooperation in the field of conservation of biological diversity and the sustainable use of biological and genetic resources. Special attention should be given to the development and strengthening of national capabilities by means of human resource development and institution-building, including the transfer of technology and/or development of research and management facilities, such as herbaria, museums, gene banks, and laboratories, related to the conservation of biodiversity;
(*d*) Without prejudice to the relevant provisions of the Convention on Biological Diversity, facilitate for this chapter the transfer of technologies relevant to the conservation of biological diversity and the sustainable use of biological resources or technologies that make use of genetic resources and cause no significant damage to the environment, in conformity with chapter 34, and recognizing that technology includes biotechnology;
(*e*) Promote cooperation between the parties to relevant international conventions and action plans with the aim of strengthening and coordinating efforts to conserve biological diversity and the sustainable use of biological resources;
(*f*) Strengthen support for international and regional instruments, programmes and action plans concerned with the conservation of biological diversity and the sustainable use of biological resources;

(*g*) Promote improved international coordination of measures for the effective conservation and management of endangered/non-pest migratory species, including appropriate levels of support for the establishment and management of protected areas in transboundary locations;

(*h*) Promote national efforts with respect to surveys, data collection, sampling and evaluation, and the maintenance of gene banks.

Means of Implementation

Financing and Cost Evaluation

The Conference secretariat has estimated the average total annual cost (1993-2000) of implementing the activities of this chapter to be about $3.5 billion, including about $1.75 billion from the international community on grant or concessional terms. These are indicative and order-of-magnitude estimates only and have not been reviewed by Governments. Actual costs and financial terms, including any that are non-concessional, will depend upon, *inter alia,* the specific strategies and programmes Governments decide upon for implementation.

Scientific and Technological Means

Specific aspects to be addressed include the need to develop:

(*a*) Efficient methodologies for baseline surveys and inventories, as well as for the systematic sampling and evaluation of biological resources;

(*b*) Methods and technologies for the conservation of biological diversity and the sustainable use of biological resources;

(*c*) Improved and diversified methods for *ex situ* conservation with a view to the long-term conservation of genetic resources of importance for research and development.

Human Resource Development

There is a need, where appropriate, to:

(*a*) Increase the number and/or make more efficient use of trained personnel in scientific and technological fields relevant to the conservation of biological diversity and the sustainable use of biological resources;

(*b*) Maintain or establish programmes for scientific and technical education and training of managers and professionals, especially in developing countries, on measures for the identification, conservation of biological diversity and the sustainable use of biological resources;

(*c*) Promote and encourage understanding of the importance of the measures required for the conservation of biological diversity and the sustainable use of biological resources at all policy-making and decision-making levels in Governments, business enterprises and lending institutions, and promote and encourage the inclusion of these topics in educational programmes.

Capacity-building

There is a need, where appropriate, to:

(*a*) Strengthen existing institutions and/or establish new ones responsible for the conservation of biological diversity and to consider the development of mechanisms such as national biodiversity institutes or centres;

(*b*) Continue to build capacity for the conservation of biological diversity and the sustainable use of biological resources in all relevant sectors;

(*c*) Build capacity, especially within Governments, business enterprises and bilateral and multilateral development agencies, for integrating biodiversity concerns, potential benefits and opportunity cost calculations into project design, implementation and evaluation processes, as well as for evaluating the impact on biological diversity of proposed development projects;

(*d*) Enhance the capacity of governmental and private institutions, at the appropriate level, responsible for protected area planning and management to undertake intersectoral coordination and planning with other governmental institutions, non-governmental organizations and, where appropriate, indigenous people and their communities.

ENVIRONMENTALLY SOUND MANAGEMENT OF BIOTECHNOLOGY

Biotechnology is the integration of the new techniques emerging from modern biotechnology with the well-established approaches of traditional biotechnology. Biotechnology, an emerging knowledge-intensive field, is a set of enabling techniques for bringing about specific man-made changes in deoxyribonucleic acid (DNA), or genetic material, in plants, animals and microbial systems, leading to useful products and technologies. By itself, biotechnology cannot resolve all the fundamental problems of environment and development, so expectations need to be tempered by realism. Nevertheless, it promises to make a significant contribution in enabling the development of, for example, better health care, enhanced food security through sustainable agricultural practices, improved supplies of potable water, more efficient industrial development processes for transforming raw materials, support for sustainable methods of afforestation and reforestation, and detoxification of hazardous wastes. Biotechnology also offers new opportunities for global partnerships, especially between the countries rich in biological resources (which include genetic resources) but lacking the expertise and investments needed to apply such resources through biotechnology and the countries that have developed the technological expertise to transform biological resources so that they serve the needs of sustainable development. Biotechnology can assist in the conservation of those resources through, for example, *ex situ* techniques. The programme areas set out below seek to foster internationally agreed principles to be applied to ensure the environmentally sound management of biotechnology, to engender public trust and confidence, to promote the development of sustainable applications of biotechnology and to establish appropriate enabling mechanisms, especially within developing countries, through the following activities:

(*a*) Increasing the availability of food, feed and renewable raw materials;

(*b*) Improving human health;
(*c*) Enhancing protection of the environment;
(*d*) Enhancing safety and developing international mechanisms for cooperation;
(*e*) Establishing enabling mechanisms for the development and the environmentally sound application of biotechnology.

Increasing the Availability of Food, Feed and Renewable Raw Materials

Basis for Action

To meet the growing consumption needs of the global population, the challenge is not only to increase food supply, but also to improve food distribution significantly while simultaneously developing more sustainable agricultural systems. Much of this increased productivity will need to take place in developing countries. It will require the successful and environmentally safe application of biotechnology in agriculture, in the environment and in human health care. Most of the investment in modern biotechnology has been in the industrialized world. Significant new investments and human resource development will be required in biotechnology, especially in the developing world.

Objectives

The following objectives are proposed, keeping in mind the need to promote the use of appropriate safety measures based on programme area:

(*a*) To increase to the optimum possible extent the yield of major crops, livestock, and aquaculture species, by using the combined resources of modern biotechnology and conventional plant/animal/micro-organism improvement, including the more diverse use of genetic material resources, both hybrid and original. Forest product yields should similarly be increased, to ensure the sustainable use of forests;
(*b*) To reduce the need for volume increases of food, feed and

raw materials by improving the nutritional value (composition) of the source crops, animals and micro-organisms, and to reduce post-harvest losses of plant and animal products;

(*c*) To increase the use of integrated pest, disease and crop management techniques to eliminate overdependence on agrochemicals, thereby encouraging environmentally sustainable agricultural practices;

(*d*) To evaluate the agricultural potential of marginal lands in comparison with other potential uses and to develop, where appropriate, systems allowing for sustainable productivity increases;

(*e*) To expand the applications of biotechnology in forestry, both for increasing yields and more efficient utilization of forest products and for improving afforestation and reforestation techniques. Efforts should be concentrated on species and products that are grown in and are of value particularly for developing countries;

(*f*) To increase the efficiency of nitrogen fixation and mineral absorption by the symbiosis of higher plants with micro-organisms;

(*g*) To improve capabilities in basic and applied sciences and in the management of complex interdisciplinary research projects.

Activities

Management-related Activities

Governments at the appropriate level, with the assistance of international and regional organizations and with the support of non-governmental organizations, the private sector and academic and scientific institutions, should improve both plant and animal breeding and micro-organisms through the use of traditional and modern biotechnologies, to enhance sustainable agricultural output to achieve food security, particularly in developing countries, with due regard to the prior identification of desired characteristics before modification, taking into account the needs of farmers, the socioeconomic, cultural and environmental impacts of modifications and the need to promote sustainable social and economic development, paying particular attention to how the use of biotechnology will impact on the maintenance of environmental integrity.

More specifically, these entities should:

(*a*) Improve productivity, nutritional quality and shelf-life of food and animal feed products, with efforts including work on pre- and post-harvest losses;

(*b*) Further develop resistance to diseases and pests;

(*c*) Develop plant cultivars tolerant and/or resistant to stress from factors such as pests and diseases and from abiotic causes;

(*d*) Promote the use of underutilized crops of possible future importance for human nutrition and industrial supply of raw materials;

(*e*) Increase the efficiency of symbiotic processes that assist sustainable agricultural production;

(*f*) Facilitate the conservation and safe exchange of plant, animal and microbial germ plasm by applying risk assessment and management procedures, including improved diagnostic techniques for detection of pests and diseases by better methods of rapid propagation;

(*g*) Develop improved diagnostic techniques and vaccines for the prevention and spread of diseases and for rapid assessment of toxins or infectious organisms in products for human use or livestock feed;

(*h*) Identify more productive strains of fast-growing trees, especially for fuel wood, and develop rapid propagation methods to aid their wider dissemination and use;

(*i*) Evaluate the use of various biotechnology techniques to improve the yields of fish, algal and other aquatic species;

(*j*) Promote sustainable agricultural output by strengthening and broadening the capacity and scope of existing research centres to achieve the necessary critical mass through encouragement and monitoring of research into the development of biological products and processes of productive and environmental value that are economically and socially feasible, while taking safety considerations into account;

(*k*) Promote the integration of appropriate and traditional biotechnologies for the purposes of cultivating genetically modified plants, rearing healthy animals and protecting forest genetic resources;

(*l*) Develop processes to increase the availability of materials derived from biotechnology for use in food, feed and renewable raw materials production.

Data and Information

The following activities should be undertaken:

(*a*) Consideration of comparative assessments of the potential of the different technologies for food production, together with a system for assessing the possible effects of biotechnologies on international trade in agricultural products;

(*b*) Examination of the implications of the withdrawal of subsidies and the possible use of other economic instruments to reflect the environmental costs associated with the unsustainable use of agrochemicals;

(*c*) Maintenance and development of data banks of information on environmental and health impacts of organisms to facilitate risk assessment;

(*d*) Acceleration of technology acquisition, transfer and adaptation by developing countries to support national activities that promote food security.

International and Regional Cooperation and Coordination

Governments at the appropriate level, with the support of relevant international and regional organizations, should promote the following activities in conformity with international agreements or arrangements on biological diversity, as appropriate:

(*a*) Cooperation on issues related to conservation of, access to and exchange of germ plasm; rights associated with intellectual property and informal innovations, including farmers' and breeders' rights; access to the benefits of biotechnology; and bio-safety;

(*b*) Promotion of collaborative research programmes, especially in developing countries, to support activities outlined in this programme area, with particular reference to cooperation with local and indigenous people and their communities in the conservation of biological diversity and sustainable use of biological resources, as well as the fostering of traditional methods and knowledge of such groups in connection with these activities;

(*c*) Acceleration of technology acquisition, transfer and adaptation by developing countries to support national activities that promote food security, through the development of systems for substantial and sustainable productivity increases that do not damage or endanger local ecosystems;

(*d*) Development of appropriate safety procedures based on programme area, taking account of ethical considerations.

Means of Implementation

Financing and Cost Evaluation

The Conference secretariat has estimated the average total annual cost (1993-2000) of implementing the activities of this programme to be about $5 billion, including about $50 million from the international community on grant or concessional terms. These are indicative and order-of-magnitude estimates only and have not been reviewed by Governments. Actual costs and financial terms, including any that are non-concessional, will depend upon, *inter alia*, the specific strategies and programmes Governments decide upon for implementation.

Human Resource Development

Training of competent professionals in the basic and applied sciences at all levels (including scientific personnel, technical staff and extension workers) is one of the most essential components of any programme of this kind. Creating awareness of the benefits and risks of biotechnology is essential. Given the importance of good management of research resources for the successful completion of large multidisciplinary projects, continuing programmes of formal training for scientists should include managerial training. Training programmes should also be developed, within the context of specific projects, to meet regional or national needs for comprehensively trained personnel capable of using advanced technology to reduce the "brain drain" from developing to developed countries. Emphasis should be given to encouraging collaboration between and training of scientists, extension workers and users to produce integrated systems. Additionally, special consideration should be given to the execution of programmes for training and exchange of knowledge on traditional biotechnologies and for training on safety procedures.

Capacity-building

Institutional upgrading or other appropriate measures will be needed to build up technical, managerial, planning and administrative capacities at the national level to support the activities in this programme area. Such measures should be backed up by international, scientific, technical and financial assistance adequate to facilitate technical cooperation and raise the capacities of the developing countries. Programme area E contains further details.

Improving Human Health

Basis for Action

The improvement of human health is one of the most important objectives of development. The deterioration of environmental quality, notably air, water and soil pollution owing to toxic chemicals, hazardous wastes, radiation and other sources, is a matter of growing concern. This degradation of the environment resulting from inadequate or inappropriate development has a direct negative effect on human health. Malnutrition, poverty, poor human settlements, lack of good-quality potable water and inadequate sanitation facilities add to the problems of communicable and non-communicable diseases. As a consequence, the health and well-being of people are exposed to increasing pressures.

Objectives

The main objective of this programme area is to contribute, through the environmentally sound application of biotechnology to an overall health programme, to:

(*a*) Reinforce or inaugurate (as a matter of urgency) programmes to help combat major communicable diseases;
(*b*) Promote good general health among people of all ages;
(*c*) Develop and improve programmes to assist in specific treatment of and protection from major non-communicable diseases;
(*d*) Develop and strengthen appropriate safety procedures based on programme area, taking account of ethical considerations;
(*e*) Create enhanced capabilities for carrying out basic and applied research and for managing interdisciplinary research.

Activities

Management-related Activities

Governments at the appropriate level, with the assistance of international and regional organizations, academic and scientific institutions, and the pharmaceutical industry, should, taking into account appropriate safety and ethical considerations:

(*a*) Develop national and international programmes for identifying and targeting those populations of the world most in need of improvement in general health and protection from diseases;
(*b*) Develop criteria for evaluating the effectiveness and the benefits and risks of the proposed activities;
(*c*) Establish and enforce screening, systematic sampling and evaluation procedures for drugs and medical technologies, with a view to barring the use of those that are unsafe for the purposes of experimentation; ensure that drugs and technologies relating to reproductive health are safe and effective and take account of ethical considerations;

(*d*) Improve, systematically sample and evaluate drinking-water quality by introducing appropriate specific measures, including diagnosis of water-borne pathogens and pollutants;

(*e*) Develop and make widely available new and improved vaccines against major communicable diseases that are efficient and safe and offer protection with a minimum number of doses, including intensifying efforts directed at the vaccines needed to combat common diseases of children;

(*f*) Develop biodegradable delivery systems for vaccines that eliminate the need for present multiple-dose schedules, facilitate better coverage of the population and reduce the costs of immunization;

(*g*) Develop effective biological control agents against disease-transmitting vectors, such as mosquitoes and resistant variants, taking account of environmental protection considerations;

(*h*) Using the tools provided by modern biotechnology, develop, *inter alia*, improved diagnostics, new drugs and improved treatments and delivery systems;

(*i*) Develop the improvement and more effective utilization of medicinal plants and other related sources;

(*j*) Develop processes to increase the availability of materials derived from biotechnology, for use in improving human health.

Data and Information

The following activities should be undertaken:

(*a*) Research to assess the comparative social, environmental and financial costs and benefits of different technologies for basic and reproductive health care within a framework of universal safety and ethical considerations;

(*b*) Development of public education programmes directed at decision makers and the general public to encourage awareness and understanding of the relative benefits and risks of modern biotechnology, according to ethical and cultural considerations.

International and Regional Cooperation and Coordination

Governments at the appropriate levels, with the support of relevant international and regional organizations, should:

(*a*) Develop and strengthen appropriate safety procedures based on programme area, taking account of ethical considerations;
(*b*) Support the development of national programmes, particularly in developing countries, for improvements in general health, especially protection from major communicable diseases, common diseases of children and disease-transmitting factors.

Means of Implementation

To achieve the above goals, the activities need to be implemented with urgency if progress towards the control of major communicable diseases is to be achieved by the beginning of the next century. The spread of some diseases to all regions of the world calls for global measures. For more localized diseases, regional or national policies will be more appropriate. The achievement of goals calls for:

(*a*) Continuous international commitment;
(*b*) National priorities with a defined time-frame;
(*c*) Scientific and financial input at global and national levels.

Financing and Cost Evaluation

The Conference secretariat has estimated the average total annual cost (1993-2000) of implementing the activities of this programme to be about $14 billion, including about $130 million from the international community on grant or concessional terms. These are indicative and order-of magnitude estimates only and have not been reviewed by Governments. Actual costs and financial terms, including any that are non-concessional, will depend upon, *inter alia*, the specific strategies and programmes Governments decide upon for implementation.

Scientific and Technological Means

Well-coordinated multidisciplinary efforts involving cooperation between scientists, financial institutions and industries will be required. At the global level, this may mean collaboration between research institutions in different countries, with funding at the intergovernmental level, possibly supported by similar collaboration at the national level. Research and development support will also need to be strengthened, together with the mechanisms for providing the transfer of relevant technology.

Human Resource Development

Training and technology transfer is needed at the global level, with regions and countries having access to, and participation in exchange of, information and expertise, particularly indigenous or traditional knowledge and related biotechnology. It is essential to create or enhance endogenous capabilities in developing countries to enable them to participate actively in the processes of biotechnology production. The training of personnel could be undertaken at three levels:

(*a*) That of scientists required for basic and product-oriented research;
(*b*) That of health personnel (to be trained in the safe use of new products) and of science managers required for complex intermultidisciplinary research;
(*c*) That of tertiary-level technical workers required for delivery in the field.

Enhancing Protection of the Environment

Basis for Action

Environmental protection is an integral component of sustainable development. The environment is threatened in all its biotic and abiotic components: animals, plants, microbes and ecosystems comprising biological diversity; water, soil and air, which form the physical components of habitats and ecosystems; and all the interactions between the components of biodiversity and their sustaining habitats and ecosystems. With the continued increase in the use of chemicals, energy and nonrenewable resources by an expanding global population, associated environmental problems will also increase. Despite increasing efforts to prevent waste accumulation and to promote recycling, the amount of environmental damage caused by overconsumption, the quantities of waste generated and the degree of unsustainable land use appear likely to continue growing.

The need for a diverse genetic pool of plant, animal and microbial germ plasm for sustainable development is well established. Biotechnology is one of many tools that can play an important role in supporting the rehabilitation of degraded ecosystems and landscapes. This may be done through the development of new techniques for reforestation and afforestation, germ plasm conservation, and cultivation of new plant varieties. Biotechnology can also contribute to the study of the effects exerted on the remaining organisms and on other organisms by organisms introduced into ecosystems.

Objectives

The aim of this programme is to prevent, halt and reverse environmental degradation through the appropriate use of biotechnology in conjunction with other technologies, while supporting safety procedures as an integral component of the programme. Specific objectives include the inauguration as soon as possible of specific programmes with specific targets:

(*a*) To adopt production processes making optimal use of natural resources, by recycling biomass, recovering energy and minimizing waste generation;

(*b*) To promote the use of biotechnologies, with emphasis on bio-remediation of land and water, waste treatment, soil conservation, reforestation, afforestation and land rehabilitation;

(*c*) To apply biotechnologies and their products to protect environmental integrity with a view to long-term ecological security.

Activities

Management-related Activities

Governments at the appropriate level, with the support of relevant international and regional organizations, the private sector, non-governmental organizations and academic and scientific institutions, should:

(*a*) Develop environmentally sound alternatives and improvements for environmentally damaging production processes;

(*b*) Develop applications to minimize the requirement for unsustainable synthetic chemical input and to maximize the use of environmentally appropriate products, including natural products;

(*c*) Develop processes to reduce waste generation, treat waste before disposal and make use of biodegradable materials;

(*d*) Develop processes to recover energy and provide renewable energy sources, animal feed and raw materials from recycling organic waste and biomass;

(*e*) Develop processes to remove pollutants from the environment, including accidental oil spills, where conventional techniques are not available or are expensive, inefficient or inadequate;

(*f*) Develop processes to increase the availability of planting materials, particularly indigenous varieties, for use in afforestation and reforestation and to improve sustainable yields from forests;

(*g*) Develop applications to increase the availability of stress-tolerant planting material for land rehabilitation and soil conservation;

(*h*) Promote the use of integrated pest management bas ed on the judicious use of bio-control agents;

(*i*) Promote the appropriate use of bio-fertilizers within national fertilizer programmes;

(*j*) Promote the use of biotechnologies relevant to the conservation and scientific study of biological diversity and the sustainable use of biological resources;

(*k*) Develop easily applicable technologies for the treatment of sewage and organic waste;
(*l*) Develop new technologies for rapid screening of organisms for useful biological properties;
(*m*) Promote new biotechnologies for tapping mineral resources in an environmentally sustainable manner.

Data and Information

Steps should be taken to increase access both to existing information about biotechnology and to facilities based on global databases.

International and Regional Cooperation and Coordination

Governments at the appropriate level, with the support of relevant international and regional organizations, should:

(*a*) Strengthen research, training and development capabilities, particularly in developing countries, to support the activities outlined in this programme area;
(*b*) Develop mechanisms for scaling up and disseminating environmentally sound biotechnologies of high environmental importance, especially in the short-term, even though those biotechnologies may have limited commercial potential;
(*c*) Enhance cooperation, including transfer of biotechnology, between participating countries for capacity-building;
(*d*) Develop appropriate safety procedures based on programme area, taking account of ethical considerations.

Means of Implementation

Financing and Cost Evaluation

The Conference secretariat has estimated the average total annual cost (1993-2000) of implementing the activities of this programme to be about $1 billion, including about $10 million from the international community on grant or concessional terms. These are indicative and order-of magnitude estimates only and have not been reviewed by Governments. Actual costs and financial terms, including any that are non-concessional, will depend upon, *inter alia*, the specific strategies and programmes Governments decide upon for implementation.

Human Resource Development

The activities for this programme area will increase the demand for trained personnel. Support for existing training programmes needs to be increased, for example, at the university and technical institute level, as well as the exchange of trained personnel between countries and regions. New and additional training programmes also need to be developed, for example, for technical and support personnel. There is also an urgent need to improve the level of understanding of biological principles and their policy implications among decision makers in Governments, and financial and other institutions.

Capacity-building

Relevant institutions will need to have the responsibility for undertaking, and the capacity (political, financial and workforce) to undertake, the above-mentioned activities and to be dynamic in response to new biotechnological developments.

Enhancing Safety and Developing International Mechanisms for Cooperation

Basis for Action

There is a need for further development of internationally agreed principles on risk assessment and management of all aspects of biotechnology, which should build upon those developed at the national level. Only when adequate and transparent safety and border-control procedures are in place will the community at large be able to derive maximum benefit from, and be in a much better position to accept the potential benefits and risks of, biotechnology. Several fundamental principles could underlie many of these safety procedures, including primary consideration of the organism, building on the principle of familiarity, applied in a flexible framework, taking into account national requirements and recognizing that the logical progression is to start with a step-by-step and case-bycase approach, but also recognizing that experience has shown that in many instances a more comprehensive approach should be used, based on the experiences of the first period, leading, *inter alia,* to streamlining and categorizing; complementary consideration of risk assessment and risk management; and classification into contained use or release to the environment.

Objectives

The aim of this programme area is to ensure safety in biotechnology development, application, exchange and transfer through international agreement on principles to be applied on risk assessment and management, with particular reference to health and environmental considerations, including the widest possible public participation and taking account of ethical considerations.

Activities

The proposed activities for this programme area call for close international cooperation. They should build upon planned or existing activities to accelerate the environmentally sound application of biotechnology, especially in developing countries.

Management-related Activities

Governments at the appropriate level, with the support of relevant international and regional organizations, the private sector, non-governmental organizations and academic and scientific institutions, should:

(*a*) Make the existing safety procedures widely available by collecting the existing information and adapting it to the specific needs of different countries and regions;

(*b*) Further develop, as necessary, the existing safety procedures to promote scientific development and categorization in the areas of risk assessment and risk management (information requirements; databases; procedures for assessing risks and conditions of release; establishment of safety conditions;

monitoring and inspections, taking account of ongoing national, regional and international initiatives and avoiding duplication wherever possible);

(*c*) Compile, update and develop compatible safety procedures into a framework of internationally agreed principles as a basis for guidelines to be applied on safety in biotechnology, including consideration of the need for and feasibility of an international agreement, and promote information exchange as a basis for further development, drawing on the work already undertaken by international or other expert bodies;

(*d*) Undertake training programmes at the national and regional levels on the application of the proposed technical guidelines;

(*e*) Assist in exchanging information about the procedures required for safe handling and risk management and about the conditions of release of the products of biotechnology, and cooperate in providing immediate assistance in cases of emergencies that may arise in conjunction with the use of biotechnology products.

International and Regional Cooperation and Coordination

Governments at the appropriate level, with the support of the relevant international and regional organizations, should raise awareness of the relative benefits and risks of biotechnology.

(*a*) Organizing one or more regional meetings between countries to identify further practical steps to facilitate international cooperation in bio-safety;

(*b*) Establishing an international network incorporating national, regional and global contact points;

(*c*) Providing direct assistance upon request through the international network, using information networks, databases and information procedures;

(*d*) Considering the need for and feasibility of internationally agreed guidelines on safety in biotechnology releases, including risk assessment and risk management, and considering studying the feasibility of guidelines which could facilitate national legislation on liability and compensation.

Means of Implementation

Financing and Cost Evaluation

The UNCED secretariat has estimated the average total annual cost (1993-2000) of implementing the activities of this programmes to be about $2 million from the international community on grant or concessional terms. These are indicative and order-of-magnitude estimates only and have not been reviewed by Governments. Actual costs and financial terms, including any that are nonconcessional, will depend upon, *inter alia*, the specific strategies and programmes Governments decide upon for implementation.

Capacity-building

Adequate international technical and financial assistance should be provided and technical cooperation to developing countries facilitated in order to build up technical, managerial, planning and administrative capacities at the national level to support the activities in this programme area.

Establishing Enabling Mechanisms for the Development and the Environmentally Sound Application of Biotechnology

Basis for Action

The accelerated development and application of biotechnologies, particularly in developing countries, will require a major effort to build up institutional capacities at the national and regional levels. In developing countries, enabling factors such as training capacity, know-how, research and development facilities and funds, industrial building capacity, capital (including venture capital) protection of intellectual property rights, and expertise in areas including marketing research, technology assessment, socio-economic assessment and safety assessment are frequently inadequate. Efforts will therefore need to be made to build up capacities in these and other areas and to match such efforts with appropriate levels of financial support. There is therefore a need to strengthen the endogenous capacities of developing countries by means of new international initiatives to support research in order to speed up the development and application of both new and conventional biotechnologies to serve the needs of sustainable development at the local, national and regional levels. National mechanisms to allow for informed comment by the public with regard to biotechnology research and application should be part of the process.

Some activities at the national, regional and global levels already address the issues outlined in programme areas A, B, C and D, as well as the provision of advice to individual countries on the development of national guidelines and systems for the implementation of those guidelines. These activities are generally uncoordinated, however, involving many different organizations, priorities, constituencies, time-scales, funding sources and resource constraints. There is a need for a much more cohesive and coordinated approach to harness available resources in the most effective manner. As with most new technologies, research in biotechnology and the application of its findings could have significant positive and negative socio-economic as well as cultural impacts. These impacts should be carefully identified in the earliest phases of the development of biotechnology in order to enable appropriate management of the consequences of transferring biotechnology.

Objectives

The objectives are as follows:

(*a*) To promote the development and application of biotechnologies, with special emphasis on developing countries, by:
   (*i*) Enhancing existing efforts at the national, regional and global levels;
   (*ii*) Providing the necessary support for biotechnology, particularly research and product development, at the national, regional and international levels;
   (*iii*) Raising public awareness regarding the relative beneficial aspects of and risks related to biotechnology, to contribute to sustainable development;
   (*iv*) Helping to create a favourable climate for investments, industrial capacity building and distribution/marketing;

(*v*) Encouraging the exchange of scientists among all countries and discouraging the "brain drain";

(*vi*) Recognizing and fostering the traditional methods and knowledge of indigenous peoples and their communities and ensuring the opportunity for their participation in the economic and commercial benefits arising from developments in biotechnology;

(*b*) To identify ways and means of enhancing current efforts, building wherever possible on existing enabling mechanisms, particularly regional, to determine the precise nature of the needs for additional initiatives, particularly in respect of developing countries, and to develop appropriate response strategies, including proposals for any new international mechanisms;

(*c*) To establish or adapt appropriate mechanisms for safety appraisal and risk assessment at the local, regional and international levels, as appropriate.

Activities

Management-related Activities

Governments at the appropriate level, with the support of international and regional organizations, the private sector, non-governmental organizations and academic and scientific institutions, should:

(*a*) Develop policies and mobilize additional resources to facilitate greater access to the new biotechnologies, particularly by and among developing countries;

(*b*) Implement programmes to create greater awareness of the potential and relative benefits and risks of the environmentally sound application of biotechnology among the public and key decision makers;

(*c*) Undertake an urgent review of existing enabling mechanisms, programmes and activities at the national, regional and global levels to identify strengths, weaknesses and gaps, and to assess the priority needs of developing countries;

(*d*) Undertake an urgent follow-up and critical review to identify ways and means of strengthening endogenous capacities within and among developing countries for the environmentally sound application of biotechnology, including, as a first step, ways to improve existing mechanisms, particularly at the regional level, and, as a subsequent step, the consideration of possible new international mechanisms, such as regional biotechnology centres;

(*e*) Develop strategic plans for overcoming targeted constraints by means of appropriate research, product development and marketing;

(*f*) Establish additional quality-assurance standards for biotechnology applications and products, where necessary.

Data and Information

The following activities should be undertaken: facilitation of access to existing information dissemination systems, especially among developing countries; improvement of such access where appropriate; and consideration of the development of a directory of information.

International and Regional Cooperation and Coordination

Governments at the appropriate level, with the assistance of international and regional organizations, should develop appropriate new initiatives to identify priority areas for research based on specific problems and facilitate access to new biotechnologies, particularly by and among developing countries, among relevant undertakings within those countries, in order to strengthen endogenous capacities and to support the building of research and institutional capacity in those countries.

Means of Implementation

Financing and Cost Evaluation

The Conference secretariat has estimated the average total annual cost (1993-2000) of implementing the activities of this programme

to be about $5 million from the international community on grant or concessional terms. These are indicative and order-of-magnitude estimates only and have not been reviewed by Governments. Actual costs and financial terms, including any that are nonconcessional, will depend upon, *inter alia*, the specific strategies and programmes Governments decide upon for implementation.

Scientific and Technological Means

Workshops, symposia, seminars and other exchanges among the scientific community at the regional and global levels, on specific priority themes, will need to be organized, making full use of the existing scientific and technological manpower in each country for bringing about such exchanges.

Human Resource Development

Personnel development needs will need to be identified and additional training programmes developed at the national, regional and global levels, especially in developing countries. These should be supported by increased training at all levels, graduate, postgraduate and post-doctoral, as well as by the training of technicians and support staff, with particular reference to the generation of trained manpower in consultant services, design, engineering and marketing research. Training programmes for lecturers training scientists and technologists in advanced research institutions in different countries throughout the world will also need to be developed, and systems giving appropriate rewards, incentives and recognition to scientists and technologists will need to be instituted. Conditions of service will also need to be improved at the national level in developing countries to encourage and nurture trained manpower with a view to retaining that manpower locally. Society should be informed of the social and cultural impact of the development and application of biotechnology.

Capacity-building

0Biotechnology research and development is undertaken both under highly sophisticated conditions and at the practical level in many countries. Efforts will be needed to ensure that the necessary infrastructure facilities for research, extension and technology activities are available on a decentralized basis. Global and regional collaboration for basic and applied research and development will also need to be further enhanced and every effort should be made to ensure that existing national and regional facilities are fully utilized. Such institutions already exist in some countries and it should be possible to make use of them for training purposes and joint research projects. Strengthening of universities, technical schools and local research institutions for the development of biotechnologies and extension services for their application will need to be developed, especially in developing countries.

Bibliography

Ager, D.V. (1963), *Principles of Palaeccolony*, McGraw-Hill, London.

——, (1976), *The Nature of Fossil record*, Proceedings of the Geologists Association.

Aldous, T., *Battle for the Environment*, London: Fontana/Collins, 1972.

Allen, J.R. (1975), *Physical Geology*, George Allen and Unwin, London.

Anderson, N., *Work and Leisure*, London: Routledge and Kegam Paul, 1961.

Arvil, R. (1967), *Man and Environment*, Crisis and the Strategy of Choice, Penguin, Hamondsworth.

Ash, M. (1972), *Planners and Ecologists*, Town and Country Planning, Vol. 40.

Ashby, W.R. (1958), *An Introduction to Cybermetics*, John Wiley and Sons, Inc., New York.

Atkinson, B.W. (1981), *Precipitation in Man and Environmental Processes* edited by K.J. Gregony and D.E. Wailling, Butterworths.

Avvill, R., *Man and Environment*, London: Penguin, 1967.

Bahuguna, S. (1990), Tehri—The Dam of Discontent, *Yojana*, Vol. 34(10), pp. 9-12.

Balsdon, J.P.V.D., *Life and Leisure in Ancient Rome*, London: Bodley Head, 1966.

Begon, W.D. and Mortimer, M. (1981), *Population Ecology*, Blackwell Oxford.

Bharucha, M.P. (1994), Environment Compliance Litigation, *Chartered Secretary*.

Bhatnagar, R.P. and V. Agrawal (1995), *Educational Adminsitration*, R. Lall Book Depot.

Billings, W.D. (1964), *"Plants and Ecosystem'*, MacMillan and Co., London.

Bird, E.C.F. (1981), *Coastal Processes in Man and Environmental Processes* edited by K.J. Gegory and D.E. Wareling, Butterworths, pp. 82-101.

Botkin, D.B. and Keller E.A. (1982), *Environmental Studies*, C.E. Merrill Publishing Company, A Bell and Howell Company, Columbus, p. 505.

—— (1982), *Environmental Studies*, C.E. Merrill Company.

Bridge, E. (1978), *World Soils*, 2nd edition, Cambridge University Press.

Brock, T.D. (1967), *The Ecosystem and Study of State*, Bioscience, Vol. 17, pp. 166-9.

Browen, H.J.M. (1966), *Trace Elements in Biochemistry*, New York, Academic Press.

Brundtland Report (1987), *Our Common Future*, Report of the World Commission on Environment and Development, Oxford University Press, p. 383.

Bryan, R.B. (1981), *Soil Erosion and Conservation in Man and Environmental Processes* edited by K.J. Georgoy and D.E. Walling, Butterworths.

Buhalis, D. and Flicher J., *Environmental Impact on Tourist Destinations: An Economic Analysis*, University of Aegean, Mytilinine, 1992.

Cain, S.A. (1971), *Foundations of Plant Geography*, Hafner Publishing Co., New York.

Cariquist, S. (1974), *Island Biology*, Columbia University Press, New York.

Carson, R. (1962), *Silent Spring*, Houghton Mifflin, New York.

Clapham, W.B. (1973), *'Natural Ecosystem'*, MacMillan, London.

Clark, G.L. (1954), *'Elements of Ecology'*, John Wiley and Sons Inc., New York.

Cloud, P.E. (1969), *'Resources and Man'*, W.H. Freeman and Company, San Francisco.

Cole, M. (1971): *Plants, Animals and Environments*, Geographical Magazine, Vol. 44, pp. 230-31.

Cosgrove, Isabel and Jackson, R., *The Geography of Recreation and Leisure*, London: Hutchinson, 1972.

Dassaman, R.D. (1976), *Environmental Conservation*, Wiley, New York.

Daubernmire, R.F. (1974), *Plants and Environment*, 3rd ed., John Wiley, New York.

Deshbandhu and G. Berberet (1987), '*Environmental Education for Conservation and Development*' Indian Env. Society, New Delhi, p. 537

Det Wyer, T.R. and Marcus, M.G. (1972), '*Urbanization and Environment*', Duxbey Press, Belmont, California.

—— (1971), *Man's Impact on Environment*, McGraw-Hill, New York.

Dikshit, R.D. (1984), '*Geography and Teaching of the Environment*'. Poona University, pp. 68-83.

Dower, M., *The Challenge of Leisure*, London: Civic Trust, 1965.

Dumazedier, J., *Towards a Society of Leisure*, New York: Free Press, 1967.

Edmunds (1973), *Environmental Administration*, New York: McGraw-Hill,

Elton, C.S. (1967), *Animal Ecology*, University of Washington Press, Washington.

—— (1989), *The Ecology of Invastion by Plants and Animals*, Methuen, London.

Embleton, C. (1989), *Natural Hazards and Global Change*, ITC Journal (1989), pp. 169-78.

Etherington, J.R. (1975), *Environment and Plant Ecology*, Wiley, New York.

Furley, P.A. and Newey, W.W. (1983), *Man and the Biosphere*, Butterworths, London.

Garg, R.K. and Prakash Tatair (1988), *Paryavan Shiksha*' Community Centre, p. 36.

Goel, M.K. (1995), *Apna Paryavaran*, Vinod Pustak Mandir, Agra, p. 314.

Goudle, A. (1984), *The Nature of the Environment*, Basil Blackwell Publisher Ltd., p. 331.

Govt. of India, (1986), *The Environment (Protection) Act, 1986*, Ministry of Environment & Forests, GOI, New Delhi, p. 14.

Gregory, K.J. and Wailing, D.E. (1981), *Man and Environment Process*, Butterworths, London.

Hewitt, K. and Hare, F.K. (1973), *Man and Environment: Conceptual Frameworks*, Commission on College Geography Resource Paper 20.

Holliman, J. (1974), *Consumer's Guide to the Protection of the Environment*, Ballanine, London.

Hutchinson, G.E. (1978), *An Introduction to Population Ecology*, Yale University Press, New Haven.

Jeffers, J.N.R. (1973), *System Modelling and Analysis in Resource Management*, Journal of Environmental Management, Vol. 1, pp. 13-28.

Kendeigh, S.C. (1974), *Ecology with Special Reference to Animal and Man*, Prentice-Hall, New Jersey.

Kerbs, C.J. (1985), *Ecology*, 3rd ed., Harper & Row, New York.

Khoshoo, T.N. (1984), *Environmental Concerns and Strategies*, Indian Environmental Society.

Komondy, E.J. (1969), *Concept of Ecology*, Prentice-Hall, New Jersey.

Kumar, V.K. (1982), *A Study in Environmental Pollution*, Tara Book Agency, Varanasi, p. 205.

Laeeq Futehally, *Our Environment*, NBT, New Delhi.

Lee N. and Wood, C. (1972), *Planning and Pollution*, The Planner, Vol. 58, pp. 153-58.

Lohani, B.N. (1984), *Environmental Quality Control*, South Asian Publishers, New Delhi, p. 448.

Madhav Gadgil and Ram Chandra Guha, *The Fissured Land; An Ecological History of India*, OUP, Delhi, 1992.

Marsh, G.P. (1984), *Man and Nature* (Physical Geography as modified by Human Action), Charles Scribner, New York.

Mash, R. (1989), *The Rights of Nature*, University of Wisconsin Press, Madison.

Mishra, R. (1968), *Ecology Work Book*, Oxford & IBH, New Delhi.

Nelson, I.G. and Byrne, A.B. (1986), *Man as an Instrument of Landscape Change,* Alberta Geog. Rev., Vol. 58, pp. 226-38.

Nicholson, M. (1972), *The Environmental Revolution,* Penguin, Harmonds-worth.

—— (1977), *The Environmental Revolution,* London: Penguin,

Odum, E.P. (1979), *Fundamental of Ecology,* Saunders, Philadelphia.

—— (1983), *Basic Ecology,* Holt Saunders Intle ed., Japan.

Odum, H.T. (1971), *Environment, Power and Society,* Wiley Interscience, New York.

Oestereich, J. (1994a), The Concept of Environmental Monitoring in the Indian Context, *Productivity,* 35(2) : 185-191.

Pal, B.P. (1981), *National Policy on Environment,* Deptt. of Environment, Government of India, New Delhi, p. 15.

Pal, S.K. and Malhotra, Sudha (1994), *Environment Trend and Thoughts in Education,* Vol. XI, 1994, Published by Innovative Research Association, Allahabad.

Paric, C.C. (1980), *Ecology and Environment,* Longmans, Londons, p. 272.

Parker, S., *The Future of Work and Leisure,* London: Mac Gibbon and Kee, 1971.

Pattern, B.C. (Ed.) 1971, *Systems Analysis and Simulation in Ecology,* Academic Press, London.

Phillipson, I. (1986), *Ecological Energetics,* Arnold, London.

Pimm, S.L. (1982), *Food Webs,* Chapman & Hall, London.

Rao, K.L. (1975), *India's Water Wealth,* Orient Longman Limited, New Delhi, p. 235.

Rao, M. Sitaram (1987), *Introduction to Social Forestry,* Oxford & IBH, p. 87.

Raymond, F. (1978), *Ecological Principles for Economic Development,* London: John Wiley.

Ricklefs, R.E. (1979), *Ecology,* 2nd ed. Chiron Press, New York.

Robinson, G.W.S., *"The Recreation Geography of South Asia",* Geographical Review, October, 1972.

Rouse, W.R. (1981), *Man-Induced Climates in Man and Environmental Process* (ed.) Walling Butterworth, pp. 38-54.

Sandbrook, R. (1992), *From Stockholm to Rio*, in *Earth Summit 92*, Regency Press Corp. London.

Sapru, R.K. (1987), *Environmental Management in India*, Ashish Publishing House, p. 288.

Saxena, A.B. (1986), *Environmental Education*, National Psychological Corporation, Agra, p. 191.

Sharma, C.B.S.R. Panneerselvam, N. and Arumikkili, K. (1985), *Environmental Genetic Toxicology in India.*

Sharma, P.D. (1990), *Ecology and Environment*, Rastogi Publishers, Meerut (U.P.).

Sharma, R.A. (1995), *Distance Education—Theory, Practice and Research*, Meerut, p. 392.

—— (1996), *Fundamental of Educational Psychology*, Surya Publication, Meerut.

Sharma, S.P. (2003), *Teacher Education—Principles Theory and Practice* Delhi, p. 574.

Sharma, Y.K. (2002), *Fundamental Aspects of Educational Technology*, Delhi, p. 454.

Simmons, I.G. (1974), *The Ecology and Natural Resources*, Edward Arnold, London.

Singh, S. (1995), *Environmental Geography*, Prayag Pustak Bhawan, Allahabad, p. 517.

—— and Dubey, A. (1989), *Environmental Management*, Geography Department, Allahabad University.

Strahler, A.N. and Strahler, A.H. (1986), *Geography and Man's Environment*, John Wiley, New York.

University of Poona (1981) *Integrated Studies in Western Ghats*, University of Poona Press.

Valiathan, M.S. (1993), *Presidential Address, 68 Annual Meeting, Association of Indian Universities*, AIU, New Delhi.

Vandana Shiva, *The Violence of the Green Revolution.*

Verma, P.S., and Agarwal, V.K. (1993), *Environmental Biology (Principles of Ecology)*, S. Chand & Company, New Delhi, p. 592.

Vernadsky, W.I. (1884), *Studies in Geochemistry*, Leningrad.

Vidya Niwas Mishra (ed.), *Creativity and Environment*, Sahitya Akademi; New Delhi, 1992.

Vyas, H. (1995), *Paryadvaran Shiksha,* Vidya Vihar, New Delhi, p. 288.

Webster's Seventh New Collegiate Dictionary, (1970), *Environment,* p. 278, G. & C. Merriam Co. Massachusets, p. 1221.

Wernes Wolfgang (ed.), *Aspects of Ecological Problems and Environmental Awareness in South Asia,* Manohar, New Delhi, 1993.

Whittaker, R.H. (1975), *Communities and Ecosystem,* 2nd ed. MacMillan, New York.

Wilson, E.C. (1975), *Sociobiology: The New Synthesis,* Harvard University Press, Cambridge, Mass.

Woodbury, A.M. (1974): *Principles of General Ecology,* McGraw Hill Book Co., New York.

Zelinsky, W. (1986), *A Prologue Population Geography,* Prentice-Hall, Englewood Cliff, New Jersey.

Index